交通类复合型人才培养专业英语丛书

机械工程专业英语

English for Mechanical Engineering

兰惠清　主编
史红梅　主审

中国铁道出版社有限公司
CHINA RAILWAY PUBLISHING HOUSE CO., LTD.

内 容 简 介

本书旨在使读者掌握机械工程专业英语术语及用法，培养并提高读者阅读和翻译专业英语文献资料的能力。内容新颖，与时俱进，较常规的机械工程专业英语书有重大突破，交通特色鲜明。主要内容结合交通应用背景，以现代机械工程为主线，包括机械工程简介、机械工程相关的热点话题、机械制造、机械设计及理论、机械电子工程、车辆工程、材料科学与技术等相关知识点，介绍了工业 4.0、高铁、机器学习等热点前沿技术。编写形式有较大创新，充分利用现代数字技术，打造立体化新形态图书，正文相关知识配英文视频、微课等拓展知识二维码。

本书适合作为交通类高等院校机械工程类专业本科生和研究生的专业英语教材，也可作为"一带一路"人才培养的培训教材，还可供涉外相关工程技术人员参考使用。

图书在版编目（CIP）数据

机械工程专业英语/兰惠清主编. —北京：中国铁道出版社有限公司，2019.8

（交通类复合型人才培养专业英语丛书）

ISBN 978-7-113-25901-3

Ⅰ.①机… Ⅱ.①兰… Ⅲ.①机械工程-英语-高等学校-教材 Ⅳ.①TH

中国版本图书馆 CIP 数据核字(2019)第 152532 号

书　　名：	机械工程专业英语
作　　者：	兰惠清

策　　划：	田银香	编辑部电话：010-63589185 转 2053		电子信箱：403195044@qq.com	
责任编辑：	田银香				
封面设计：	刘　颖				
责任校对：	张玉华				
责任印制：	郭向伟				

出版发行：中国铁道出版社有限公司（100054，北京市西城区右安门西街 8 号）

网　　址：http://www.tdpress.com/51eds/

印　　刷：三河市航远印刷有限公司

版　　次：2019 年 8 月第 1 版　2019 年 8 月第 1 次印刷

开　　本：850 mm×1 168 mm　1/16　印张：18.75　字数：396 千

书　　号：ISBN 978-7-113-25901-3

定　　价：48.00 元

版权所有　侵权必究

凡购买铁道版图书，如有印制质量问题，请与本社教材图书营销部联系调换。电话：(010) 63550836

打击盗版举报电话：(010) 51873659

前　言

据统计，全球有超过40%的人在学习和使用英语；超过85%的科技资料（如书籍、期刊、专利说明书和内部技术报告等）运用英语语言；不论在哪个国家召开国际学术会议，所规定使用的工作语言一般也都是英语。因此，英语既是人们交流的主要工具，也是开展科学研究的重要工具。

伴随"一带一路"的建设和中国高铁"走出去"战略的实施，海外高铁项目建设、装备制造等国际交流合作愈发紧密，这就要求机械工程类专业学生既要精通技术知识，又要熟练应用专业英语，要成为具备国际技术交流能力的复合型人才。

目前，各高校纷纷开设专业英语课程。本书的编写是为满足机械工程专业英语课程教学需要和涉外人员的自学参考要求。内容结合高铁等交通应用背景，以现代机械工程为主线，借助英语工具开展机械工程在交通领域的应用实践，而不是通过专业实践去巩固英语知识，最终要提升读者运用英语进行交通背景下机械工程专业活动的整体水平，能够进行国际交流研讨、专业英语文本写作、了解国际科技创新、在国外顺畅进行施工交流等实践，使其将专业能力、英语能力与应用能力三者有机融合。

本书共分7章，包括机械工程简介、机械工程相关的热点话题、机械制造、机械设计及理论、机械电子工程、车辆工程、材料科学与技术等相关知识点。内容较常规的机械工程专业英语书都有重大突破，交通特色鲜明。首先，内容新颖、与时俱进，介绍工业4.0、高铁、机器学习等热点前沿技术；其次，覆盖面广、体系完整，涵盖机械工程一级学科下所有的二级研究方向，尤其是车辆工程为独立一章；第三，学科知识与交通背景结合，例如，介绍钢轨打磨、钢轨在线监测、轨道交通材料、复合材料等内容。另外，编写形式有较大创新，充分利用现代数字技术，打造立体化新形态图书，正文相关知识配二维码，添加用英文讲解的机械工程视频、前沿技术英文微课等拓展素材。

本书由北京交通大学的相关专家和技术人员编写。兰惠清任主编，负责全书的构思、组稿和统稿。各章编写分工如下：第1章和第2章由兰惠清编写，第3章由曹建国、房善想和张勤俭编写，第4章由张朝辉编写，第5章由董立静和陈光荣编写，第6章由陈星宇编写，第7章由杜云慧编写。全书由史红梅主审，李聪协助完成统稿工作。

在本书的编写过程中，编者参考了国内外专家学者的最新研究成果和资料，在此向这些专家学者表示诚挚的感谢。

机械工程所涉及的内容广泛，学科跨度大，新材料、新技术发展迅速，加之编者水平有限，书中难免有不足之处，恳请广大读者提出宝贵意见。

编　者
2019年5月

CONTENTS

目　　录

Chapter 1　Introduction of Mechanical Engineering
　　　　　　机械工程简介 1
　1.1　Definition and History of Mechanical Engineering
　　　　（机械工程的定义及其发展历史）... 1
　1.2　Fields of Mechanical Engineering（机械工程的分类）.................................... 4
　1.3　Advantages of Mechanical Engineering in China（我国机械工程的优势）.... 6
　1.4　Prospects of Mechanical Engineering in China（我国机械工程的前景）..... 16

Chapter 2　Hot Topics About Mechanical Engineering
　　　　　　机械工程相关的热点话题 18
　2.1　Industry 4.0（工业 4.0）... 18
　2.2　Made in China 2025（中国制造 2025）... 22
　2.3　High Speed Railway in China（中国高铁）... 27
　2.4　Machine Learning（机器学习）... 30

Chapter 3　Mechanical Manufacturing
　　　　　　机械制造 37
　3.1　CAD/CAM/CAPP（计算机辅助设计/制造/工艺规划）................................. 37
　3.2　Flexible Manufacturing System（柔性制造系统）.. 48
　3.3　Computer Numerical Control（计算机数字控制）.. 54
　3.4　Rail Grinding（钢轨打磨）... 61
　3.5　Micromachine and Nanomachine（微纳米加工）.. 67

Chapter 4　Mechanical Design and Theory
　　　　　　机械设计及理论 78
　4.1　Problems in Mechanical Design（机械设计问题）.. 78
　4.2　Machine Elements（机械零件）... 84
　4.3　Friction, Wear and Lubrication（摩擦、磨损和润滑）................................. 97

4.4　Industrial Robot（机器人） ……………………………………………………… 105

Chapter 5　Mechatronic Engineering
　　　　　　机械电子工程 ………………………………………………………………… 112
　　5.1　Advanced Control and Automation of Mechatronic Systems
　　　　（机电系统的先进控制与自动化）…………………………………………… 112
　　5.2　On-Line Monitoring and Fault Diagnosis of Mechatronic Systems
　　　　（机电系统的在线监测与故障诊断）………………………………………… 136
　　5.3　Fluid Transmission and Control（流体传动与控制）……………………… 144
　　5.4　Embedded System and Intelligent Instrument
　　　　（嵌入式系统与智能仪器）…………………………………………………… 172

Chapter 6　Vehicle Engineering
　　　　　　车辆工程 ……………………………………………………………………… 185
　　6.1　Vehicle Structure Reliability（车辆结构可靠性）………………………… 185
　　6.2　Vehicle System Dynamics and Control（车辆系统动力学与控制）……… 191
　　6.3　Vehicle Vibration and Noise Control（车辆振动与噪声控制）…………… 200
　　6.4　Vehicle Digital Design（车辆数字化设计）………………………………… 210

Chapter 7　Materials Science and Technology
　　　　　　材料科学与技术 ……………………………………………………………… 218
　　7.1　Materials for Rail Traffic（轨道交通材料）………………………………… 218
　　7.2　Ceramic Material and Composite Material（陶瓷与复合材料）………… 239
　　7.3　Brake Disc of Multiple Units（动车组制动盘）…………………………… 258
　　7.4　Liquid Die Forging and Semi-Solid Forming（液态模锻和半固态成形）… 270

References
参考文献 ……………………………………………………………………………………… 290

Chapter 1
Introduction of Mechanical Engineering

1.1 Definition and History of Mechanical Engineering

1.1.1 Definition of Mechanical Engineering

Mechanical engineering (ME) is the discipline that applies engineering, physics, engineering mathematics, and materials science principles to design, analyze, manufacture, and maintain mechanical systems. It is one of the oldest and broadest of the engineering disciplines.

Definition of mechanical engineering

The mechanical engineering field requires an understanding of core areas including mechanics, dynamics, thermodynamics, materials science, structural analysis, and electricity. In addition to these core principles, mechanical engineers use tools such as computer-aided design (CAD), computer-aided manufacturing (CAM), and product life cycle management to design and analyze manufacturing plants, industrial equipment and machines, heating and cooling systems, transport systems, aircraft, watercraft, robotics, medical devices, weapons, and others. It is the branch of engineering that involves the design, production, and operation of machinery.

1.1.2 History of Mechanical Engineering

The application of mechanical engineering can be seen in the archives of various ancient and medieval societies. In ancient Greece, the works of Archimedes influenced mechanics in the Western tradition and Hero of Alexandria created the steam engine—Aeolipile. In China, Zhang Heng improved a water clock and invented a seismometer, and Ma Jun invented a chariot with differential gears. The medieval Chinese horologist and engineer Su Song incorporated an escapement mechanism into his astronomical clock tower two centuries before escapement devices were found in medieval European clocks. He also invented the world's first known endless power-transmitting chain drive.

During the 7th to the 15th century, some inventors made remarkable contributions in the field of mechanical technology. Al-Jazari, who was one of them, wrote *Book of Knowledge of Ingenious Mechanical Devices* in 1206 and presented many mechanical designs. He is also considered to be the inventor of such mechanical devices which now form the very basic of mechanisms, such as the crankshaft and camshaft.

History of mechanical engineering

During the 17th century, important breakthroughs in the foundations of mechanical engineering occurred in England. Sir Isaac Newton formulated Newton's Laws of Motion and developed Calculus, the mathematical basis of physics. Newton was reluctant to publish his works for years, but he was finally persuaded to do so by his colleagues, such as Sir Edmond Halley, much to the benefit of all mankind. Gottfried Wilhelm Leibniz is also credited with creating Calculus during this time period.

During the early 19th century industrial revolution, machine tools were developed in England, Germany, and Scotland. This allowed mechanical engineering to develop as a separate field within engineering. They brought manufacturing machines and the engines to power them. The first British professional society of mechanical engineers Institution of Mechanical Engineers was formed in 1847; thirty years after the civil engineers formed the first such professional society Institution of Civil Engineers. On the European continent, Johann von Zimmermann (1820–1901) founded the first factory for grinding machines in Chemnitz, Germany in 1848.

In the United States, the American Society of Mechanical Engineers (ASME) was formed in 1880, becoming the third such professional engineering society, after the American Society of Civil Engineers in 1852 and the American Institute of Mining Engineers in 1871. The first schools in the United States to offer an engineering education were the United States Military Academy in 1817, an institution now known as Norwich University in 1819, and Rensselaer Polytechnic Institute in 1825. Education in mechanical engineering has historically been based on a strong foundation in mathematics and science.

Words and Expressions

mechanics [mɪˈkænɪks] the branch of physics that deals with the action of forces on material objects with mass	*n.* 力学（这里指静力学）

dynamics [daɪˈnæmɪks] the branch of mechanics that is concerned with the effects of forces on the motion of objects	n. 动力学
thermodynamics [ˌθɜːməʊdaɪˈnæmɪks] the science of the conversions between heat and other forms of energy	n. 热力学
computer-aided design (CAD) [kəmˈpjuːtə eɪdɪd dɪˈzaɪn]	计算机辅助设计
computer-aided manufacturing (CAM) [kəmˈpjuːtə eɪdɪd ˌmænjuˈfæktʃərɪŋ]	计算机辅助制造
product life cycle management [ˈprɒdʌkt laɪf ˈsaɪkl ˈmænɪdʒmənt]	产品生命周期管理；全生命周期管理
aircraft [ˈeəkrɑːft] a vehicle capable of atmospheric flight due to interaction with the air, such as buoyancy or lift	n. 航空器
watercraft [ˈwɔːtəkrɑːft] a boat, ship, sea scooter, or similar vehicle	n. 船舶
seismometer [saɪzˈmɒmɪtə] a device used by seismologists to detect and measure seismic waves and therefore locate earthquakes	n. 测震仪
chariot [ˈtʃærɪət] a two-wheeled horse-drawn vehicle, used in Bronze Age and Early Iron Age warfare	n. 双轮敞篷马车（古代用于战争）
horologist [ˈhɒrɒlədʒɪst] someone who makes or repairs watches or clocks	n. 时计学者
escapement mechanism [ɪˈskeɪpmənt ˈmekənɪzəm]	n. 擒纵机构
endless power-transmitting chain drive [ˈendləs ˈpaʊə trænsˈmɪtɪŋ tʃeɪn draɪv]	无级动力传动链传动
American Society of Mechanical Engineers (ASME) [əˈmerɪkən səˈsaɪəti əv məˈkænɪkl ˌendʒɪˈnɪəz]	美国机械工程师协会

Questions for Discussion

1. Give a definition of mechanical engineering.
2. What branches does mechanical engineering mainly include?
3. What functions does mechanical engineering have?

1.2 Fields of Mechanical Engineering

Fields of mechanical engineering

The discipline of mechanical engineering is the science of the theory, method and technology of the performance, design and manufacture of mechanical systems and products, including two major fields of mechanics and manufacturing science.

Mechanics is a science that studies the mechanical structure and system performance and its design theory and method, including the mechanism, transmission, dynamics, strength, tribology, design, bionics, micro mechanics and interface mechanics involved in the manufacturing process and the mechanical system.

Manufacturing science is a science about manufacturing processes and systems. It covers product design, forming and manufacturing (casting, plastic forming, joint forming, mold manufacturing, surface engineering, etc.), processing and manufacturing (ultra-precision machining, efficient processing, non-traditional machining, complex surface processing, measurement and instrument, equipment design and manufacturing, surface functional structure manufacturing, micromanufacturing, nanomanufacturing, biomimetic and raw material production), and operation and management of the manufacturing system and other science.

According to Ministry of Education of the People's Republic of China, there are four classical second-class disciplines belong to the first-class discipline of Mechanical Engineering: mechanical manufacture and automation, mechanical design and theory, mechatronic engineering, and vehicle engineering.

1.2.1 Mechanical Manufacture and Automation

Mechanical manufacture and automation is one of the engineering disciplines studying mechanical manufacturing theory, manufacturing technology, automated manufacturing system and advanced manufacturing mode. The discipline combines the latest development of related disciplines, making manufacturing technology, manufacturing systems and manufacturing models showing a new look. The goal of mechanical manufacture and automation is very clear, which combines mechanical equipment and automation through the way of computer to form a series of advanced manufacturing technologies, including CAD (computer aided design), CAM (computer aided manufacturing), FMS (flexible manufacturing system) and so on. Finally, large-scale computer integrated manufacturing system (CIMS) is formed, which makes the traditional machining process get a qualitative leap. Specific applications in industry include computer numerical control (CNC) machine tools, machining centers and so on.

1.2.2 Mechanical Design and Theory

Mechanical design and theory is a basic technical discipline to analyze, synthesize, quantitatively describe and control the performance of the machinery. It is a brief introduction to the detailed work processes and procedures of the mechanical engineering. Therefore, work principles, motion and dynamic properties, strength and life, vibration and noise, friction, wear and lubrication, mechanical innovation and design, and modern design calculation methods of various machines, mechanisms and parts are mainly studied.

1.2.3 Mechatronic Engineering

Mechanical and electronic engineering, commonly known as mechatronics, is a kind of mechanical engineering and automation. The majors of mechanical and electronic engineering include basic theoretical knowledge and mechanical design and manufacturing methods, the application ability of computer software and hardware, and the ability to design, manufacture, testing and developing various kinds of mechanical and electrical products and systems. Mechanical and electronic engineering is the product of high-speed development of science and technology and interlinking of disciplines. It breaks the traditional classification of disciplines and integrates many technical characteristics. It represents the emergence of new technologies, new ideas, new research methods and new research objectives.

1.2.4 Vehicle Engineering

Vehicle engineering is one of engineering technology field studying theory, design and manufacturing technology of mobile machines on the land, such as automobiles, tractors, locomotives, military vehicles and other engineering vehicles.

Vehicles are widely used in modern society. They are related to the revitalization and development of the automobile industry and transportation industry, one of the pillar industries of China's economic construction, and has a great influence on the modernization of agriculture and the modernization of national defense equipment. From the early stages of vehicle engineering, it involves mechanics, mechanical design, material, fluid mechanics, chemical industry, and extends to the interpenetrating and interrelated subjects such as mechatronic engineering, mechanical design and theory, computer, electronic technology, measurement technology, control technology and other disciplines, and further touches on wide fields such as: medicine, physiology and psychology, forming a comprehensive discipline and engineering technology covering a variety of new and high technologies.

According to the characteristics of the industry, this field covers design and manufacture of automobiles and tractors, military vehicles, locomotive vehicles, engineering vehicles, energy power and so on.

According to the working nature of engineering technicians, the scope of the field can be divided into: research and development of vehicles; manufacturing and processing of vehicles; performance testing, testing and analysis of vehicles; use, management and maintenance of vehicles, equipment related to production testing vehicles, the development of testing apparatus, etc.

Words and Expressions

mechanical manufacture and automation [məˈkænɪkl ˌmænjuˈfæktʃə ænd ɔːtəˈmeɪʃən]	机械制造及自动化
mechanical design and theory [məˈkænɪkl dɪˈzaɪn ænd ˈθɪəri]	机械设计及理论
mechatronic engineering [məkæˈtrɒnɪk endʒɪˈnɪərɪŋ]	机械电子工程
vehicle engineering [ˈviːəkl endʒɪˈnɪərɪŋ]	车辆工程

Questions for Discussion

What field of mechanical engineering attracts you the most? Why?

1.3 Advantages of Mechanical Engineering in China

In recent years, a series of outstanding progresses and original innovations have been achieved in the fields of mechanical engineering, which provide a large number of novel theories, technologies and methodologies for our country's economic construction and mechanical engineering.

1.3.1 Tribology

Tribology

Tsinghua University has made important progress in the field of nanotribology and its technology. In the study of superfine chemical mechanical polishing on the surface of the computer hard disk, the behavior mechanism of super fine surface nanoparticles was put forward. The equilibrium law of chemical and mechanical action was found, and a new technology and advanced polishing process of the superfinishing surface of the hard disk substrate were explored. The roughness of the surface wave after polishing were lower than that of nanoscale. Combined with the study of wheel-rail interaction in high speed railway, Southwest Jiaotong University reproduced the wheel-rail corrugation for the first time, and analyzed the formation mechanism of wheel-rail corrugation from the theory and test. Lanzhou Institute of Chemical Physics has played an important role in the research of nanosolid lubrication technology applied to Chinese aerospace engineering. Tribology has become one of the most influential subjects in Chinese mechanical engineering discipline in the international academia.

1.3.2 Robotics

Yanshan University, Shanghai Jiaotong University, etc., put forward a universal method and the general formula for calculating the degree of freedom of less freedom and parallel structure, and the theory of principal helix analytic recognition model with the helix theory, Lie group, and set theory as mathematical tools[1]. Moreover, using the above theory dozens of new mechanisms, force sensors, micromanipulation robots, and earthquake simulators had been developed. Tianjin University and Tsinghua University put forward a universal modeling method of the Jacobi matrix based on a linear space theory, and developed a large number of engineering equipment, such as the large gantry hybrid machine tool, the high-speed packaging robot and so on.

Robotics

1.3.3 Mechanical Dynamics

Nonlinear dynamics, fault prediction and intelligent maintenance of complex mechatronic systems are frontier research fields of mechanical dynamics. Northeastern University has proposed theory of probability and screen with constant bed thickness, vibration synchronization and control synchronization theory, and has designed dozens of engineering vibration machines. Nanjing University of Aeronautics & Astronautics proposed a control system dynamics with time delay, a bifurcation mechanism and control method of the vibration control system with elastic constraints, as well as modeling and control method of the hysteresis damping vibration control system. American control experts evaluate the system as "a refreshing system method". A coupling dynamics model of locomotive vehicle and track system has been developed by Southwest Jiaotong University, and the dynamic simulation system of locomotive vehicle and track coupling dynamics and the evaluation system of safety field test are developed.

1.3.4 Mechanical Transmission

Ultrasonic Motor Research Center of Nanjing University of Aeronautics & Astronautics has put forward theory and design method of movement mechanism, electromechanical coupling model, drive and control technology for a new ultrasonic motor. Moreover, it has invented dozens of ultrasonic motors and drivers with unique traveling wave and standing wave. In the high-speed and ultra-precision motion control research, Huazhong University of Science and Technology found and clarified the mechanism of the cyclonic phenomena of air bearings. Chongqing University invented a water-lubricated rubber alloy bearing with the organic combination of multi-surface and linear circular arc groove. This bearing saved a lot of precious metals and has been widely used in the transmission system of ships at home and abroad.

1.3.5 Biomimetic Machinery and Biological Manufacturing

Biomimetic machinery and biological manufacturing

Jilin University has made important progress in the research of bionic flexible dynamic drag reduction and biomimetic electroosmotic desorption theory. It has created and developed a mechanical bionics discipline and invented a series of ground mechanical desorption and drag reduction bionics technology, and successfully applied it to agricultural machinery and national defense engineering. In the study of artificial bone biomimetic manufacturing, Xi'an Jiaotong University established a model of bone tissue, proposed a composite structure repair method of bone defect, and made the structural frame of artificial bone by rapid prototyping, and succeeded in repairing the bone defect of the animal.

1.3.6 Advanced Electronic Manufacturing

Central South University has put forward a concept of "extreme manufacturing". The research on the key scientific problems in hard disk drive and chip manufacturing is carried out by Shanghai Jiaotong University and Tsinghua University. A single abrasive grinding method with controllable nanoscratch depth and length is proposed. A critical depth model of the grinding wheel in self-rotating grinding of silicon wafer has been established to reveal the wide frequency of the high acceleration motion system. The design theory and control method of high acceleration, high precision and high reliability precision driving platform are put forward. The rapid diffusion mechanism of atoms at ultrasonic bonding interface is clarified, the "stick-slip" motion characteristic of bonding interface is discovered, and the variable parameter loading process is put forward.

1.3.7 Digital Manufacturing

A new method of geometric reasoning based on visual cone and the unified discrimination of the contour error of complex surfaces have been put forward by Huazhong University of Science and Technology. The software system of digital modeling and manufacturability analysis for complex products is developed, and a system platform of integrated rapid measurement, digital modeling and manufacturing oriented design is established. It is applied to rapid development of complex curved surface parts, such as cylinder heads and blades. Wuhan University of Technology had put forward a digital manufacturing modeling theory, digital manufacturing resource sharing based on manufacturing grid, a theory model of agile supply chain under digital manufacturing environment, and a theory and algorithm of intelligent scheduling in digital manufacturing workshop. Moreover, it had established a remote operation, monitoring and diagnosis platform of virtual NC machining system equipment under digital manufacturing environment. Shanghai Jiaotong University had applied a distance function and pseudo-distance function theory into a qualitative and quantitative geometric reasoning of force and motion spin space. Therefore, a qualitative and quantitative analysis and evaluation index system for the sealing and stability of clamp and holding mechanism can be established.

1.3.8 Mechanical Measurement

Tianjin University has invented an on-site calibration method and device for space dimension measurement, which has solved the problem of on-site calibration and its installation urgently needed in modern manufacturing. Tsinghua University has invented a "laser nanometers" with frequency difference greater than birefringent dual-frequency lasers and displacement measurements. Harbin Institute of Technology has invented a high performance series of straight line and rotary motion datum, and a series of confocal scanning measuring devices and microscopes, which made the level and vertical resolution reach the sub nanometer scale for China to develop the first cylindrical and micro deep hole measuring instrument standard installation, so that China has ability to disseminate and trace the value in this field [2]. Chongqing University has put forward a concept and principle of "intelligent virtual control", established a unified model of signal transformation, and developed thousands of unique virtual instrument development systems. Chongqing Institute of Technology put forward an idea of "time-space transformation" of the precision displacement measurement, and invented a time-grating displacement sensor and its testing system.

1.3.9 Processing and Manufacturing

Dalian University of Technology has put forward a precision manufacturing technology and equipment of hard and brittle material complex curved radome. In view of special requirements of the electrical performance of the radome, a precision grinding theory for compensating the electrical performance of the radome is put forward. A theoretical model of the relation between the comprehensive electrical performance error of the radome and the compensation of the geometric parameters is established. The digital repair equipment has been invented, which solved a major scientific and technological problem in the defense engineering. In the research field of high-speed precision grinding, Hunan University has put forward "four point constant line speed method", which makes the grinding defects improved and the surface quality is obviously improved. Huazhong University of Science and Technology proposed a formation mechanism, theoretical model, parameter optimization and

Piston manufacturing process

control strategy of grinding surface burn, and solved the problem of grinding burn.

1.3.10　Ultra-Precision Machining

Precision machining

Harbin Institute of Technology have been studied deeply in the study of the machining mechanism of micro- and nano-cutting process, the mechanism of tool wear and tear and the mechanism of ultra-precision cutting removal of brittle materials. Many special equipment for ultra-precision cutting are developed successfully, and the ultra-precision machining of the key parts of the laser nuclear fusion key parts has been used. National University of Defense Technology has firstly broken through an ion beam and magnetorheological optical polishing technology in China, and established the basic theory of magnetic fluid and ion beam and other controllable flexible body medium polishing. A complete set of process routes and equipment for optical mirror full waveband error control have been formed, and the grade processing precision of the surface, spherical and aspherical mirror surfaces can be realized steadily in a nanoscale.

1.3.11　Design

Zhejiang University has deeply studied a product concept design and virtual prototyping technology based on Intelligent Computing, and put forward a key design technology of mass customization, which is product configuration, product variant, product evolution and product recursion, and realized the virtual simulation test of the creative generation and design performance of product design concept. The technology and system of computer aided product innovation design have been developed and docked with data interface of famous foreign systems. Hebei University of Technology has developed creatively a theory, a qualitative and quantitative analysis methods of multi-conflict and the domain transformation technology are put forward, and the creation pattern of innovative ideas in the fuzzy front end of product innovation is summarized.

1.3.12　Forming Manufacturing

Forming manufacturing

Rapid prototyping manufacturing is the most representative manufacturing technology of traditional manufacturing to multi-disciplinary and digital modern manufacturing. Xi'an Jiaotong University, Tsinghua University, and Huazhong University of Science and Technology have made a deep study of stereo lithography apparatus, laminated object manufacturing, fused deposition modeling and selected laser sintering, and promoted the formation and development of rapid prototyping. The technology has applied to the fields of automobile, medical rehabilitation engineering, agricultural water-saving devices and so on, and promoted the industrialization of new technology.

1.3.13 Microelectronic Fabrication

Northwestern Polytechnical University has proposed an integrated design tool supporting arbitrary processes. Peking University has developed three sets of standard process flow and established a high-level silicon processing platform. The first forming technology of multilayer silicon micromechanical structure, the fabrication of silicon based optical waveguide and the new method of wafer level packaging have been invented by Shanghai Institute of Microsystem and Information Technology, Chinese Academy of Sciences. And the fabrication process of the double side micromechanical piezoresistive sensor based on single silicon wafer structure is formed. Dalian University of Technology has developed an automatic manufacturing equipment for plastic microfluidic chip. It has mastered the key technology of batch production of microstructure hot pressing metal mould and microfluidic chip. A nanoscale imprinting process of "pressure preserving-release-soliding" under the normal temperature soft imprinting was put forward by Xi'an Jiaotong University. The influence of the properties of the corrosion inhibitor and the liquid-solid interface and solid-solid interface on the filling quality and the release effect of the mold cavity was found, and the nanoscale imprint was realized with a good compound fidelity. Institute of Physics, Chinese Academy of Sciences has successfully developed a probe for dual probe scanning tunneling microscope with symmetrical mechanical structure. North University of China has developed a wafer level microstructural stress testing platform based on Raman spectroscopy, and has completed static stress and dynamic stress testing.

Microelectronic fabrication

Words and Expressions

nanotribology [ˌnænəʊtraɪˈbɒlədʒi] nano+tribology, the branch of tribology that studies friction, wear, adhesion and lubrication phenomena at the nanoscale, where atomic interactions and quantum effects are not negligible	*n.* 纳米摩擦学
chemical mechanical polishing [ˈkemɪkl məˈkænɪkl ˈpɒlɪʃɪŋ]	化学机械抛光
wheel-rail interaction [wiːl reɪl ˌɪntərˈækʃən]	轮轨关系
corrugation [ˌkɒrʊˈɡeɪʃən] the process of corrugating; contraction into wrinkles or alternate ridges and grooves	*n.* 钢轨的波浪形磨耗，简称波磨
helix theory [ˈhiːlɪks ˈθɪəri]	螺旋理论
Lie group [liː ɡruːp]	李群

set theory [set ˈθɪəri]	集合论
large gantry hybrid machine tool [lɑːdʒ ˈɡæntri ˈhaɪbrɪd məˈʃiːn tuːl]	大型龙门混联机床
packaging robot [ˈpækɪdʒɪŋ ˈrəʊbɒt]	包装机器人
nonlinear dynamics [nɒnˈlɪniə daɪˈnæmɪks]	非线性动力学
fault prediction [fɔːlt prɪˈdɪkʃən]	故障预示
intelligent maintenance [ɪnˈtelɪdʒənt ˈmeɪntənəns]	智能维护
screen with constant bed thickness [skriːn wɪð ˈkɒnstənt bed ˈθɪknəs]	等厚筛分
time delay [taɪm dɪˈleɪ]	时滞
bifurcation mechanism [baɪfəˈkeɪʃən ˈmekənɪzəm]	分叉机理
coupling dynamics model of locomotive vehicle and track system [ˈkʌplɪŋ daɪˈnæmɪks ˈmɒdl əv ləʊkəˈməʊtɪv ˈviːəkl ænd træk ˈsɪstəm]	机车车辆-轨道系统耦合动力学模型
mechanical transmission [məˈkænɪkl trænzˈmɪʃən]	机械传动
ultrasonic motor [ʌltrəˈsɒnɪk ˈməʊtə(r)]	超声电机
electromechanical coupling model [ɪˌlektrəʊməˈkænɪkl ˈkʌplɪŋ mɒdl]	机电耦合模型
drive and control technology [draɪv ænd kənˈtrəʊl tekˈnɒlədʒi]	驱动与控制技术
traveling wave [ˈtrævlɪŋ weɪv]	行波
standing wave [ˈstændɪŋ weɪv]	驻波
air bearing [eə ˈbeərɪŋ]	气浮轴承
cyclonic phenomena [ˈsaɪkləʊnɪk fəˈnɒmɪnə]	气旋现象
water-lubricated [ˈwɔːtə ˈluːbrɪkeɪtɪd]	水润滑的
alloy bearing [ˈælɔɪ ˈbeərɪŋ]	合金轴承
precious metal [ˈpreʃəs ˈmetl]	贵重金属
biomimetic machinery [ˌbaɪəʊmɪˈmetɪk məˈʃiːnəri]	仿生机械
biological manufacturing [ˌbaɪəˈlɒdʒɪkl ˌmænjuˈfæktʃərɪŋ]	生物制造
bionic flexible dynamic drag reduction [baɪˈɒnɪk ˈfleksəbl daɪˈnæmɪk dræɡ rɪˈdʌkʃən]	仿生柔性动态减阻

biomimetic electroosmotic desorption theory [ˌbaɪəʊmɪˈmetɪk ɪˈlektrəʊzˈmɒtɪk dɪˈsɔːpʃən ˈθɪəri]	仿生电渗脱附理论
mechanical bionics discipline [məˈkænɪkl baɪˈɒnɪks ˈdɪsɪplɪn]	机械仿生学科
national defense engineering [ˈnæʃənl dɪˈfens endʒɪˈnɪərɪŋ]	国防工程
rapid prototyping [ˈræpɪd ˈprəʊtətaɪpɪŋ]	快速成型法
advanced electronic manufacturing [ədˈvɑːnst ɪˌlekˈtrɒnɪk ˌmænjuˈfæktʃərɪŋ]	先进电子制造
extreme manufacturing [ɪkˈstriːm ˌmænjuˈfæktʃərɪŋ]	极端制造
hard disk drive [hɑːd dɪsk draɪv]	硬盘驱动器
chip manufacturing [tʃɪp ˌmænjuˈfæktʃərɪŋ]	芯片制造
nanoscratch depth and length [ˈnænəʊskrætʃ depθ ænd leŋθ]	纳米量级划痕深度和长度
single abrasive grinding method [ˈsɪŋgl əˈbreɪsɪv ˈgraɪndɪŋ meθəd]	单颗磨粒磨削方法
silicon wafer [ˈsɪlɪkən ˈweɪfə(r)]	硅片
wide frequency [waɪd ˈfriːkwənsi]	宽频
ultrasonic bonding interface [ˌʌltrəˈsɒnɪk ˈbɒndɪŋ ˈɪntəfeɪs]	超声键合界面
rapid diffusion mechanism [ˈræpɪd dɪˈfjuːʒən ˈmekənɪzəm]	快速扩散机理
stick-slip [stɪk slɪp]	黏滑
visual cone [ˈvɪʒuəl kəʊn]	可视锥
geometric reasoning [ˌdʒɪəˈmetrɪk ˈriːzənɪŋ]	几何推理
contour error [ˈkɒntʊə ˈerə(r)]	轮廓误差
unified discrimination [ˈjuːnɪfaɪd dɪˌskrɪmɪˈneɪʃən]	统一判别
integrated rapid measurement [ˈɪntɪgreɪtɪd ˈræpɪd ˈmeʒəmənt]	集成快速测量
digital modeling [ˈdɪdʒɪtl ˈmɒdəlɪŋ]	数字建模
manufacturing oriented design [ˌmænjuˈfæktʃərɪŋ ˈɔːrɪəntɪd dɪˈzaɪn]	面向制造设计
cylinder head [ˈsɪlɪndə hed]	缸盖

blade [bleɪd] the flat functional end of a propeller, oar, hockey stick, screwdriver, skate, etc.	n. 叶片
manufacturing grid [ˌmænjuˈfæktʃərɪŋ ɡrɪd]	制造网格
digital manufacturing resource sharing [ˈdɪdʒɪtl ˌmænjuˈfæktʃərɪŋ rɪˈsɔːs ˈʃeərɪŋ]	数字制造资源共享
agile supply chain [ˈædʒaɪl səˈplaɪ tʃeɪn]	敏捷供应链
intelligent scheduling [ɪnˈtelɪdʒənt ˈskedʒuːlɪŋ]	智能调度
virtual NC machining system [ˈvɜːtʃuəl en siː məˈʃiːnɪŋ ˈsɪstəm]	虚拟数控加工系统
pseudo-distance function [ˈsjuːdəʊ ˈdɪstəns ˈfʌŋkʃən]	伪距离函数
spin space [spɪn speɪs]	旋量空间
clamp and holding mechanism [klæmp ænd ˈhəʊldɪŋ ˈmekənɪzəm]	夹具和夹持机构
mechanical measurement [məˈkænɪkl ˈmeʒəmənt]	机械测量
on-site calibration [ɒn saɪt kælɪˈbreɪʃən]	现场校准
laser nanometer [ˈleɪzə ˈnænəʊmiːtə(r)]	激光器纳米测尺
birefringent dual-frequency laser [ˌbaɪrɪˈfrɪndʒənt ˈdjuːəl ˈfriːkwənsi ˈleɪzə(r)]	双折射双频激光器
intelligent virtual control [ɪnˈtelɪdʒənt ˈvɜːtʃuəl kənˈtrəʊl]	智能虚拟控件
virtual instrument [ˈvɜːtʃuəl ˈɪnstrəmənt]	虚拟仪器
time-space transformation [taɪm speɪs trænsfəˈmeɪʃən]	时空转换
time-grating displacement sensor [taɪm ˈɡreɪtɪŋ dɪsˈpleɪsmənt ˈsensə]	时栅位移传感器
processing and manufacturing [prəˈsesɪŋ ænd ˌmænjuˈfæktʃərɪŋ]	加工制造
complex curved surface [ˈkɒmpleks kɜːvd ˈsɜːfɪs]	复杂曲面
radome [ˈreɪdəʊm] blend of radar + dome, a radar dome	n. 天线罩
four point constant line speed method [fɔː pɔɪnt ˈkɒnstənt laɪn spiːd ˈmeθəd]	四点恒线速法

parameter optimization [pəˈræmɪtə ˌɒptɪmaɪˈzeɪʃən]	参数优化
control strategy [kənˈtrəʊl ˈstrætədʒi]	控制策略
grinding burn [ˈɡraɪndɪŋ bɜːn]	磨削烧伤
Intelligent Computing [ɪnˈtelɪdʒənt kəmˈpjuːtɪŋ]	智能计算
virtual prototyping technology [ˈvɜːtʃuəl prəʊtəˈtaɪpɪŋ tekˈnɒlədʒi]	虚拟样机技术
virtual simulation [ˈvɜːtʃuəl ˌsɪmjuˈleɪʃən]	虚拟仿真
computer aided product innovation design [kəmˈpjuːtə eɪdɪd ˈprɒdʌkt ˌɪnəˈveɪʃən dɪˈzaɪn]	计算机辅助产品创新设计
data interface [ˈdeɪtə ˈɪntəfeɪs]	数据接口
fuzzy [ˈfʌzi] uncertain	*adj.* 模糊的
ultra-precision machining [ˈʌltrə prɪˈsɪʒən məˈʃiːnɪŋ]	超精密加工
ultra-precision cutting removal [ˈʌltrə prɪˈsɪʒən ˈkʌtɪŋ rɪˈmuːvəl]	超精密切削去除
ion beam [ˈaɪən biːm]	离子束
magnetorheological [mæɡˈniːtəʊˌriəˈlɒdʒɪkəl] describing a substance whose rheological properties are modified by a magnetic field	*adj.* 磁流变的
controllable flexible body medium polishing [kənˈtrəʊləbl ˈfleksɪbl ˈbɒdi ˈmiːdiəm ˈpɒlɪʃɪŋ]	可控柔体介质抛光
aspherical [æzˈferɪkl] not (quite) spherical	*adj.* 非球面的
forming manufacturing [ˈfɔːmɪŋ ˌmænjʊˈfæktʃərɪŋ]	成形制造
stereo lithography apparatus [ˈsteriəʊ lɪˈθɒɡrəfi ˌæpəˈreɪtəs]	立体光固化成形
laminated object manufacturing [ˈlæmɪneɪtɪd ˈɒbdʒɪkt ˌmænjʊˈfæktʃərɪŋ]	薄材制造
fused deposition modeling [fjuzd ˌdepəˈzɪʃən ˈmɒdəlɪŋ]	丝材熔覆
selected laser sintering [sɪˈlektɪd ˈleɪzə ˈsɪntərɪŋ]	粉材激光烧结
medical rehabilitation engineering [ˈmedɪkəl ˈriːhəˌbɪlɪˈteɪʃən endʒɪˈnɪərɪŋ]	医疗康复工程

nanoscale imprinting [ˈnænəʊskeɪl ɪmˈprɪntɪŋ]	纳米压印
corrosion inhibitor [kəˈrəʊʒən ɪnˈhɪbɪtə]	阻蚀剂

Notes

[1] 本句可译为：燕山大学、上海交通大学等院校以螺旋理论、李群、集合论等为数学工具，提出少自由度并联机结构综合的普适性方法和通用的自由度计算公式，以及主螺旋解析识别模型理论。

[2] 本句可译为：哈尔滨工业大学发明了高性能系列直线及回转运动基准装置，还发明了多种共焦扫描测量装置和显微镜，使水平、垂直分辨力达到了亚纳米量级，为我国研制出第一台圆柱度和微小深孔测量仪标准装置，使我国具备了在该领域进行量值传递和溯源的能力。

Questions for Discussion

What aspects of mechanical engineering in China are keeping ahead in the world?

1.4　Prospects of Mechanical Engineering in China

Prospects of mechanical engineering in China

Some of the previous research have played great significance at home and abroad. In particularly, a few advanced subjects have gained their top-level position in international academic community. However, it should be soberly aware that Chinese mechanical engineering science is generally in a state of backward position.

It is mainly reflected that the theory, method and technology of Chinese mechanical engineering have no significant contribution to the independent innovation and development of Chinese manufacturing industry. There are few new concepts and theories in the field of mechanical engineering in China, and few theories, methods and techniques of mechanical engineering have important international influence. Furthermore, few Chinese scholars have great influence in the field of international mechanical engineering. In general, the position of Chinese mechanical engineering in the world is lagging behind that of Chinese manufacturing industry in the international manufacturing sector.

The development of mechanical engineering in the future will mainly be restricted and promoted by two aspects. One is the innovation and development of the manufacturing industry, and the other is the evolution of the discipline. In view of the general trends of the future development of manufacturing industry (i.e. globalization, informatization, greening, knowledge-creating and extremalization), the basic task of mechanical engineering science is to provide the new theory, new method and advanced manufacturing technology for the future development of manufacturing industry.

With the progress of the world, the needs of the country and the development of disciplines, the following characteristics and trends have emerged in the development of mechanical engineering science. On the one hand, the development of high technology field, such as optoelectronics, micro- and nano- technology, aerospace, biomedicine, and major engineering technology, requires mechanical and manufacturing science to provide more and better new theories, new methods and new technologies to these fields. Then put forward the emergence and development of new fields of manufacturing science, such as micro/nano fabrication, bionics and bio manufacturing, and microelectronics manufacturing. On the other hand, along with the intersection of mechanical and manufacturing science, information science, life science, material science, management science and nanoscience and technology, besides the development of mechanism, tribology, dynamics, structural strength, transmission and design had been proceeded; bionic mechanics, nanotribology, and manufacturing information science, manufacturing management science and other new cross science are also produced and developed.

As our country will vigorously promote the advanced instruments and equipment technology with independent intellectual property rights in the future, the fundamental research on the design and manufacture of high technology instruments and equipment based on independent innovation will be paid more attention and more quickly developed. In addition, as China's resources and environment are facing unprecedented challenges in the century, it is required that the protection of the environment, the safety and green degree of products, the saving of materials and energy, the remanufacturing of mechanical and electrical equipment and the basic research of the new energy manufacturing domain are required more than ever before.

Questions for Discussion

What are the factors that restrict the development of Chinese mechanical engineering?

Chapter 2
Hot Topics About Mechanical Engineering

2.1 Industry 4.0

Industry 4.0

Industry 4.0 is a name for the current trend of automation and data exchange in manufacturing technologies. It includes cyber-physical systems, the Internet of Things (IoT), cloud computing and cognitive computing. Industry 4.0 is commonly referred to as the fourth industrial revolution.

First came steam and the first machines that mechanized some of the work our ancestors did [1]. Next was electricity, the assembly line and the birth of mass production. The third era of industry [2] came about with the advent of computers and the beginnings of automation, when robots and machines began to replace human workers on those assembly lines.

And now we enter Industry 4.0, in which computers and automation will come together in an entirely new way, with robotics connected remotely to computer systems equipped with machine learning algorithms that can learn and control the robotics with very little input from human operators.

Industry 4.0 introduces what has been called the "smart factory", in which cyber-physical systems monitor the physical processes of the factory and make decentralized decisions. The physical systems become Internet of Things, communicating and cooperating both with each other and with humans in real time via the wireless web.

There are four design principles in Industry 4.0. These principles support companies in identifying and implementing Industry 4.0 scenarios.

- Interoperability: The ability of machines, devices, sensors, and people to connect and communicate with each other via the Internet of Things or the Internet of People (IoP).
- Information transparency: The ability of information systems to create a virtual copy of the physical world by enriching digital plant models with sensor data. This requires the aggregation of raw sensor data to higher-value context information.
- Technical assistance: First, the ability of assistance systems to support humans by aggregating and visualizing information comprehensively for making informed decisions and solving urgent problems on short notice. Second, the ability of cyber physical systems to physically support humans by conducting a range of tasks that are unpleasant, too exhausting, or unsafe for their human co-workers.
- Decentralized decisions: The ability of cyber physical systems to make decisions on their own and to perform their tasks as autonomously as possible. Only in the case of exceptions, interferences, or conflicting goals, are tasks delegated to a higher level.

Current usage of the term has been criticised as essentially meaningless, in particular on the grounds that technological innovation is continuous and the concept of a "revolution" in technology innovation is based on a lack of knowledge of the details.

The characteristics given for the German government's Industry 4.0 strategy are: the strong customization of products under the conditions of highly flexible (mass-) production or mass production. The required automation technology is improved by the introduction of methods of self-optimization, self-configuration, self-diagnosis, cognition and intelligent support of workers in their increasingly complex work. The largest project in Industry 4.0 as of July 2013 is the BMBF[3] leading-edge cluster "Intelligent Technical Systems Ostwestfalen-Lippe (it's OWL)". Another major project is the BMBF project RES-COM, as well as the Cluster of Excellence "Integrative Production Technology for High-Wage Countries". In 2015, the European Commission started the international Horizon 2020 research project CREMA (Providing Cloud-based Rapid Elastic Manufacturing Based on the XaaS and Cloud Model) as a major initiative to foster the Industry 4.0 topic.

Industry 4.0 and change

Challenges in implementation of Industry 4.0 as follows:
- IT security issues, which are greatly aggravated by the inherent need to open up those previously closed production shops;
- Reliability and stability needed for critical machine-to-machine communication (M2M), including very short and stable latency times;
- Need to maintain the integrity of production processes;
- Need to avoid any IT snags, as those would cause expensive production outages;
- Need to protect industrial know-how (contained also in the control files for the industrial

automation gear);
- Lack of adequate skill-sets to expedite the march towards fourth industrial revolution;
- Threat of redundancy of the corporate IT department;
- General reluctance to change by stakeholders;
- Loss of many jobs to automatic processes and IT-controlled processes, especially for lower educated parts of society;
- Low top management commitment;
- Unclear legal issues and data security;
- Unclear economic benefits/ Excessive investment;
- Lack of regulation, standard and forms of certifications;
- Insufficient qualification of employees.

Modern information and communication technologies like cyber-physical system, big data analytics and cloud computing, will help early detection of defects and production failures, thus enabling their prevention and increasing productivity, quality, and agility benefits that have significant competitive value.

Big data analytics consists of 6Cs in the integrated Industry 4.0 and cyber physical systems environment. The 6C system comprises:
- Connection (sensor and networks);
- Cloud (computing and data on demand);
- Cyber (model & memory);
- Content/context (meaning and correlation);
- Community (sharing & collaboration);
- Customization (personalization and value).

What is big data

In this scenario and in order to provide useful insight to the factory management, data has to be processed with advanced tools (analytics and algorithms) to generate meaningful information. Considering the presence of visible and invisible issues in an industrial factory, the information generation algorithm has to be capable of detecting and addressing invisible issues such as machine degradation, component wear, etc. in the factory floor.

From both strategic and technological perspectives, the Industry 4.0 roadmap visualizes every further step on the route towards an entirely digital enterprise. In order to achieve success in the digital transformation process, it is necessary to prepare the technology roadmap in the most accurate way. In today's business, Industry 4.0 is driven by digital transformation in vertical/horizontal value chains and product/service offerings of the companies. The required key technologies for Industry 4.0 transformation such as artificial intelligence, internet of things, machine learning, cloud systems, cybersecurity, adaptive robotics cause radical changes in the business processes of organizations.

Words and Expressions

Industry 4.0 [ˈɪndəstri fɔːpɔɪnt ˈzɪərəʊ]	工业 4.0
cyber-physical system [saɪbə ˈfɪzɪkl ˈsɪstəm]	信息物理系统
Internet of Things [ˈɪntənet əv ˈθɪŋz]	物联网
Internet of People [ˈɪntənet əv ˈpiːpl]	人际网络
cloud computing [klaʊd kəmˈpjuːtɪŋ]	云计算
cognitive computing [ˈkɒɡnɪtɪv kəmˈpjuːtɪŋ]	认知计算
the fourth industrial revolution [ðə fɔːθ ɪnˈdʌstriəl ˌrevəˈluːʃən]	第四次工业革命
assembly line [əˈsembli laɪn]	装配线；流水作业线
machine learning algorithm [məˈʃiːnɪŋ ˈlɜːnɪŋ ˈælɡərɪðəm]	机器学习算法；机器学习演算法
know-how [nəʊ haʊ]	专门知识；技术诀窍
smart factory [smɑːt ˈfæktəri]	智慧工厂
interoperability [ˌɪntərˌɒpərəˈbɪləti] the capability of a product or system, to interact and function with others reciprocally	*n.* 互操作性；互用性
transparency [trænˈspærənsi] (signal processing) sufficient accuracy to make the compressed result perceptually indistinguishable from the uncompressed input	*n.* 透明；透明度
contextualize information [kənˌtekstʃuəlaɪz ˌɪnfəˈmeɪʃən]	符合情境的信息
decentralized [ˌdiːˈsentrəlaɪzd] not centralized; having no center or several centers	*adj.* 分散的
machine-to-machine communication (M2M) [məˈʃiːn tu: məˈʃiːn kəˌmjuːnɪˈkeɪʃən]	机对机通信
latency [ˈleɪtənsi] (electronics) a delay, a period between the initiation of something and the occurrence	*n.* 延迟
snag [snæɡ] (figuratively) a problem or difficulty with something	*n.* 困难

outage [ˈaʊtɪdʒ] the amount of something lost in storage or transportation	n. 供应中断
expedite [ˈekspədaɪt] (transitive) to accelerate the progress	v. 加快进展；迅速完成
stakeholder [ˈsteɪkˌhəʊldə] a person or organisation with a legitimate interest in a given situation, action or enterprise	n. 利益相关者
big data [bɪɡ ˈdeɪtə]	大数据
cybersecurity [saɪbəsɪˈkjʊərɪti] security against electronic attacks such as cyberwarfare	n. 网络安全

Notes

[1] 本句是倒装句，指工业 1.0 是机械制造时代，全句可译为：首先通过蒸汽机和首批机器使我们祖先的部分工作实现机械化。

[2] the third era of industry 工业 3.0 时代。

[3] BMBF：Geman Federal Ministry of Education and Research，德语缩写为 BMBF。

Questions for Discussion

1. Why everyone must know Industry 4.0?
2. What challenges does Industry 4.0 have?
3. Do you have any plan to work in the future with Industry 4.0?

2.2 Made in China 2025

Authorities plan

Made in China 2025 (MiC 2025) is a strategic plan of China issued in May 2015. The Center for Strategic and International Studies describes it as an "initiative to comprehensively upgrade Chinese industry" directly inspired by the German Industry 4.0. It is an attempt to move the country's manufacturing up the value chain. The goals include increasing the domestic content of core materials to 40% by 2020 and 70% by 2025. The plan focuses on high-tech fields including the pharmaceutical industry which are presently the purview of foreign companies.

2.2.1 Progress of Main Activities

(1) Plans were vigorously implemented and "Five Major Programs" yielded initial results

Ministry of Industry and Information Technology (MIIT), in collaboration with relevant agencies including National Development and Reform Commission, Ministry of Science and Technology and Ministry of Education, published and implemented eleven supporting plans and a full range of supporting policies and measures. Five major programs were carried out, including the construction of national manufacturing innovation centers, smart manufacturing, consolidation of industrial foundation, green manufacturing, and high-end equipment innovation[1]. As of now, the first innovation center—National Power Battery Innovation Center, was established, whereas preparation for establishing the National Additive Manufacturing Innovation Center was underway, and 19 provincial manufacturing innovation centers were built. 226 projects for comprehensive and standardized trial as well as new model application of smart manufacturing were implemented, 109 smart manufacturing pilot programs were selected and the first batch of 19 public service platforms for industrial technologies was established. Breakthroughs were achieved for a full range of core spare parts, key basic materials and advanced basic technique, as a result of the implementation of a series of major breakthrough actions and one package application plan. The building of a green manufacturing system was initiated. 99 enterprises were organized to conduct pilot work on green design. 51 national low-carbon industrial parks were built. 57 high-risk pollutant reduction programs were undertaken. The energy consumption per unit of the added value for industrial enterprises above the state designated scale decreased by about 5%. High-end equipment innovations began to bear fruits, with significant breakthroughs made in areas such as key and auxiliary technologies of high speed railway, and key materials for hard-shell cell of electric vehicle.

Geoeconomic impact of Made in China 2025

(2) Initial progress was made in key milestone projects focusing on major objectives

Among the 15 key milestone projects of the year, 7 were fully implemented, 4 were basically implemented, and others were in progress. Some projects achieved breakthroughs. For example, the first flexible hybrid industrial robot was successfully developed, and the capability for its small-volume production achieved. RV reducers has the capability for volume production with overseas orders and harmonic wave decelerator is capable of large-scale production.

(3) Implementation of pilot projects has taken root

The work of setting up pilot cities / city clusters for "Made in China 2025" was actively promoted, with 8 cities (Ningbo, Quanzhou, Shenyang, Changchun, Wuhan, Wuzhong, Qingdao and Chengdu) approved as pilot cities, and five cities in south Jiangsu province, west bank of Pearl River in Guangdong Province and Changsha, Zhuzhou and Xiangtan of Hunan Province approved as pilot city clusters. Currently, in line with the implementation plan, all pilot cities /city clusters started comprehensive pilot work on industrial upgrading, technological innovation, policy guarantee and talent training, and began to explore new models and paths for manufacturing transformation and upgrading under the new normal situation with a view to gain experiences that can be replicated elsewhere and applied widely as early as possible [2].

(4) Region-specific guidelines were introduced to facilitate differentiated development across regions

Following the principle of "promoting differentiated development and ministry-province cooperation based on comparative advantages", MIIT strengthened alignment with all localities and formulated "Province and Municipality Specific Guidelines for Made in China 2025 (2016)" to help solve such long-standing problems as redundant construction and homogeneous competition. In line with "Made in China 2025", and taking into account of local resource advantages, industrial development status, geographical conditions, and market environment, all localities identified development priorities and goals and developed related policies and measures to accelerate industrial transformation and upgrading[3]. By the end of 2016, 29 provinces, autonomous regions and municipalities have promulgated implementation plans or guiding documents designed to pursue a new environment for developing manufacturing, which suits local conditions, boasts distinctive local features, and ensures coordination across regions and differentiated competition.

(5) The integration of manufacturing industry and Internet achieved noted results thanks to more effective policy guidance

Guided by policies such as Guidelines on Promoting Internet + and Guidelines on Deepening Integration of Manufacturing Industry and Internet as issued by the State Council and as a result of efforts by all parties concerned, Internet was widely applied in the R&D and designing process. The key products and equipment became more intelligent. In 2016, the penetration rate of applying digital R&D tools for enterprises reached 61.8% and the digitalization of key processes reached 33.3%. The development of "mass entrepreneurship and innovation" platforms for large enterprises yielded good results, in which, 47% established and operated platforms for collaborative innovation and the state-owned key enterprises built up 110 Internet-based platforms for "mass entrepreneurship and innovation". Standard implementation of the administrative system for the integration of manufacturing industry and internet proceeded smoothly involving over 4,000 enterprises nationwide. More than 72,000 enterprises did self-assessment, self-diagnosis, and self-benchmarking, with 8.8% decrease of operating cost and a 6.9% increase of operating profit on average for these enterprises. Industrial Internet was promoted and prospective studies focusing on top-level architecture and key standards as priorities were undertaken. Collaboration among enterprises, academia, research institutes was strengthened and application was accelerated through pilot programs[4].

(6) Initial results were achieved in quality improvement and branding

Since the implementation of the special action on enhancing the variety, quality and branding for consumer goods industry, the General Office of the State Council issued Opinions on Special Actions of Enhancing the Variety, Quality and Brand for Consumer Goods Industry and Creating Favorable Market Environment and Plan for Improving Standard and Quality of Consumer Goods (2016-2020), and MIIT, General Administration of Quality Supervision, Inspection and Quarantine, and State Administration of Science, Technology and Industry for National Defense jointly released jointly Guidelines on Special Actions of Enhancing Quality and Brand of Equipment Manufacturing Industry and Plan for Enhancing Standardization and Quality Improvement of Equipment Manufacturing Industry. In 2016, 23 quality control and technical evaluation labs for industrial products were ratified, 669 compulsory standards and plans were merged and streamlined, and 208 industrial standards for consumer goods were released. 11 international standard drafts were proposed, which laid a solid basis for improving the quality of consumer goods. 251 pilot enterprises for brand development were identified across the industry. The first batch of 22 industrial clusters participating in the pilot projects for regional brand development enjoyed on average 2.3% market increase, with over 10% export growth rate and the output value ratio of new products increased from 27% to 34.1%.

2.2.2 Main Difficulties and Problems

(1) International environment remains complicated and challenging

The global economy is now in a period of moving toward new growth drivers, and the role of traditional engines to drive growth has weakened. Despite the emergence of new technologies such as artificial intelligence and 3D printing, new sources of growth are yet to emerge, and the global economy has remained sluggish. Anti-globalization sentiment and protectionism are on the rise, bringing more uncertainties.

(2) Innovative development path remains to be explored

As there is no off-the-shelf experience to draw from [5] in terms of transforming and upgrading the manufacturing sector and achieving the shift between new and old driving forces, we have to rid ourselves of dependence on traditional ways of development. There is still a long way to go when dealing with issues such a show to cultivate new industry and transform the old one, stabilize growth and adjust structure, and to address the balance of government guidance and market role.

(3) Market forces need to be allowed to play a more effective role

Faced with pressure from economic downturns, some regions and sectors are in a rush to promote investment and stabilize growth. Meanwhile, we need to be fully aware of issues like the decrease of investment, particularly private investment in the manufacturing sector, and the extreme pressure that some manufacturing companies face in production and operation. Necessary policies and measures must be developed promptly to provide effective guidance.

Words and Expressions

Made in China 2025 (MiC 2025) [meɪd ɪn ˈtʃaɪnə tuː ˈθaʊznd ænd ˈtwenti faɪv]	中国制造 2025
pharmaceutical [ˌfɑːməˈsuːtɪkl] (medicine) of, or relating to pharmacy or pharmacists	*adj.* 制药的；配药的
purview [ˈpɜːvjuː] scope or range of interest or control	*n.* 范围
Ministry of Industry and Information Technology [ˈmɪnɪstri əv ˈɪndəstri ænd ɪnfəˈmeɪʃən tekˈnɒlədʒi]	工业和信息化部
National Development and Reform Commission [ˈnæʃnəl dɪˈveləpmənt ænd ˈrɪˈfɔːm kəˈmɪʃən]	国家发展和改革委员会
National Additive Manufacturing Innovation Center [ˈnæʃnəl ˈædɪtɪv mænjuˈfæktʃərɪŋ ɪnəʊˈveɪʃən ˈsentə]	国家增材制造创新中心
pilot city [ˈpaɪlət ˈsɪti]	试点城市
breakthrough [ˈbreɪkθruː] any major progress; such as a great innovation or discovery that overcomes a significant obstacle	*n.* 突破行动
one package application plan [wʌn ˈpækɪdʒ ˌæplɪˈkeɪʃən plæn]	"一条龙"应用计划
low-carbon industrial park [ləʊ ˈkɑːbən ɪnˈdʌstriəl pɑːk]	低碳工业园区
high-risk pollutant reduction program [ˈhaɪ ˈrɪsk pəˈluːtənt rɪˈdʌkʃən ˈprəʊɡræm]	高风险污染物削减项目
milestone project [ˈmaɪlstəʊn ˈprɒdʒekt]	标志性项目
flexible hybrid industrial robot [ˈfleksəbl ˈhaɪbrɪd ɪnˈdʌstriəl ˈrəʊbɒt]	柔性复合工业机器人
R&D and designing process [dɪˈzaɪnɪŋ ˈprəʊses]	研发设计环节
General Administration of Quality Supervision, Inspection and Quarantine [ˈdʒenərəl ədˌmɪnɪsˈtreɪʃən əv ˈkwɒlɪti sjuːpəˈvɪʒən ɪnˈspekʃ ænd ˈkwɒrəntiːn]	国家质量监督检验检疫局

State Administration of Science, Technology and Industry for National Defense [steɪt ədˌmɪnɪsˈtreɪʃən əv ˈsaɪəns tekˈnɒlədʒi ænd ˈɪndəstri fɔː ˈnæʃənəl dɪˈfens]	国家国防科技工业局
sluggish [ˈslʌgɪʃ] slow; having little motion	*adj.* 行动迟缓的；反应慢的
anti-globalization sentiment [ˈænti ˌgləʊbəlaɪˈzeɪʃən ˈsentɪmənt]	"逆全球化"思潮
economic downturn [ˌiːkəˈnɒmɪk ˈdaʊntɜːn]	经济下行

Notes

[1] 本句可译为：组织实施了国家制造业创新中心建设、智能制造、工业强基、绿色制造、高端装备创新五项重大工程。

[2] 本句可译为：目前，各试点示范城市（群）按照实施方案的要求，围绕构建产业升级、科技创新、政策保障、人才培养等体系进行综合试点，积极探索新常态下制造业转型升级的新模式和新路径，力争尽快形成可复制、可推广的经验。

[3] 本句可译为：各地积极对接《中国制造2025》，根据自身资源优势、产业基础、区位条件和市场环境，选择发展重点和方向，制定相关政策措施，加快促进产业转型升级。

[4] 本句可译为：加强产学研协同，通过试点示范加快应用推广。

[5] there is no off-the-shelf experience to draw from 可译为：没有现成经验可以照搬。

Questions for Discussion

1. Why did Chinese government issue MiC2025 plan?
2. What results had "Five Major Programs" yielded?
3. What difficulties must be faced for MiC2025 in China?

2.3 High Speed Railway in China

High speed railway (HSR) in China is the country's network of passenger-dedicated railways designed for speeds of 250～350 km/h. In 2018 HSR exceeded 29,000 km in total length, accounting for about two-thirds of the world's high speed railway tracks in commercial service. It is the world's longest HSR network and is also the most extensively used.

High speed railway is officially defined as "newly-built passenger-dedicated rail lines designed for electric multiple unit (EMU) train sets traveling at not less than 250 km/h (including lines with reserved capacity for upgrade to the 250 km/h standard), on which initial service operate at not less than 200 km/h". High speed trains are defined as EMU train sets with not more than 16 railcars, axle load not greater than 17 tones and service interval of not less than three minutes. High speed railway service thus requires high speed EMU trains to run on high speed railway track at speeds of not less than 200 km/h. Passenger rail service on high speed EMU trains operating on non-high speed track or on high speed track but at speeds below 200 km/h are not considered high speed railway. Certain mixed use freight and passenger rail lines, that can be upgraded for train speeds of 250 km/h, with current passenger service of at least 200 km/h, are also considered high speed railway.

Almost all HSR trains, track and service are owned and operated by the China Railway Corporation under the brand China Railway High-speed (CRH) with only few exceptions. The China Railway High-speed (CRH) high speed train service was introduced in April 2007 featuring high speed train sets called Hexie Hao (shown in Figure 2.1) and Fuxing Hao running at speed from 250 km/h to 350 km/h on upgraded/dedicated high speed track. The Beijing-Tianjin Intercity Rail,

Figure 2.1 CRH2 Hexie Hao

which opened in August 2008 and could carry high speed trains at 350 km/h (217 m/h), was the first passenger dedicated HSR line.

China's new standardized 350 kph train-set

China developing again

China's early high speed trains were imported or built under technology transfer agreements with foreign train-makers including Alstom, Siemens, Bombardier and Kawasaki Heavy Industries. Since the initial technological support, Chinese engineers have re-designed internal train components and built indigenous trains manufactured by the state-owned CRRC Corporation.

The advent of high speed railway in China has greatly reduced travel time and has transformed Chinese society and economy. A World Bank study found "a broad range of travelers of different income levels choose HSR for its comfort, convenience, safety and punctuality".

Notable HSR lines in China include the Beijing–Hong Kong West Kowloon High Speed Railway which is the world's longest HSR line in operation, the Beijing–Shanghai high speed railway with the world's fastest operating conventional train services and the Shanghai Maglev, the world's first high speed commercial magnetic levitation line, whose trains run on non-conventional track and reach a top speed of 430 km/h.

In common parlance, high speed train service in China is often referred to by the Chinese terms "gaotie" or "dongche" which correspond to, respectively, to G-class and D-class passenger service.

- G-class (for gaotie or "high speed railway") train service generally features EMU trains running on passenger-dedicated high speed railway lines and operating at top speeds of at least 250 km/h. For example, the G7 train from Beijing South to Shanghai Hongqiao, which runs on the Beijing–Shanghai HSR, a line with designed speed of 350 km/h (217 mph), makes four stops and averages 300 km/h (186 mph) for the trip.
- D-class (for dongche or "electric multiple unit") train service features EMU trains running at lower speeds, whether on high speed or non-high speed track. D-class trains can vary widely in actual trip speed. The non-stop D211 train from Guiyang East to Guangzhou South on the Guiyang–Guangzhou HSR, a line with designed speed of 250 km/h, averages 207 km/h for the trip. The D312 EMU sleeper train between Beijing South and Shanghai on the non-high speed Beijing–Shanghai Railway averages 121 km/h for the trip.
- Certain C-class (for chengji or "intercity") train service that operate on high speed track at speeds above 250 km/h are also considered high speed railway service. For example, C-class trains on the Beijing–Tianjin ICR, a line with designed speed of 350 km/h, reach top speeds of 330 km/h (205 mph) and average 226 km/h (140 mph) for the trip.

Words and Expressions

high speed railway (HSR) [haɪ spiːd ˈreɪlweɪ] a type of rail transport that operates significantly faster than traditional rail traffic, using an integrated system of specialized rolling stock and dedicated tracks	高铁
electric multiple unit (EMU) [ɪˈlektrɪk ˈmʌltɪpl ˈjuːnɪt]	动车组
intercity [ˌɪntəˈsɪti] that connects cities with other cities	*adj.* 城际的
indigenous [ɪnˈdɪdʒɪnəs] of or relating to the native inhabitants of a land	*adj.* 本地的
parlance [ˈpɑːləns]	*n.* 用语；说法

Questions for Discussion

1. Which HSR have you taken? Why did you choose to take HSR?
2. Why is HSR in China famous in the world? Did you feel pride of the HSR technology in China?
3. What is the maximum speed of CSR in China?

2.4 Machine Learning

Machine learning is a subset of artificial intelligence in the field of computer science that often uses statistical techniques to give computers the ability to "learn" (i.e., progressively improve performance on a specific task) with data, without being explicitly programmed.

What is machine learning

The name machine learning was coined in 1959 by Arthur Samuel. Evolved from the study of pattern recognition and computational learning theory in artificial intelligence, machine learning explores the study and construction of algorithms that can learn from and make predictions on data – such algorithms overcome following strictly static program instructions by making data-driven predictions or decisions, through building a model from sample inputs. Machine learning is employed in a range of computing tasks where designing and programming explicit algorithms with good performance is difficult or infeasible; example applications include email filtering, detection of network intruders or malicious insiders working towards a data breach, optical character recognition (OCR), learning to rank, and computer vision.

Types of algorithms in machine learning

Machine learning is closely related to (and often overlaps with) computational statistics, which also focuses on prediction-making through the use of computers. It has strong ties to mathematical optimization, which delivers methods, theory and application domains to the field. Machine learning is sometimes conflated with data mining, where the latter subfield focuses more on exploratory data analysis and is known as unsupervised learning. Machine learning can also be unsupervised and be used to learn and establish baseline behavioral profiles for various entities and then used to find meaningful anomalies.

Within the field of data analytics, machine learning is a method used to devise complex models and algorithms that lend themselves to prediction; in commercial use, this is known as predictive analytics. These analytical models allow researchers, data scientists, engineers, and analysts to "produce reliable, repeatable decisions and results" and uncover "hidden insights" through learning from historical relationships and trends in the data.

2.4.1 Machine Learning Tasks

Machine learning tasks are typically classified into two broad categories, depending on whether there is a learning "signal" or "feedback" available to a learning system, specific as follows:

- Supervised learning: The computer is presented with example inputs and their desired outputs, given by a "teacher", and the goal is to learn a general rule that maps inputs to outputs. As special cases, the input signal can be only partially available, or restricted to special feedback.
- Semi-supervised learning: the computer is given only an incomplete training signal: a training set with some (often many) of the target outputs missing.
- Active learning: the computer can only obtain training labels for a limited set of instances (based on a budget), and also has to optimize its choice of objects to acquire labels for. When used interactively, these can be presented to the user for labeling.
- Reinforcement learning: training data (in form of rewards and punishments) is given only as feedback to the program's actions in a dynamic environment, such as driving a vehicle or playing a game against an opponent.
- Unsupervised learning: No labels are given to the learning algorithm, leaving it on its own to find structure in its input. Unsupervised learning can be a goal in itself (discovering hidden patterns in data) or a means towards an end (feature learning).

How machines learn

2.4.2 Machine Learning Applications

Another categorization of machine learning tasks arises when one considers the desired output of a machine-learned system, specific as follows:

- In classification, inputs are divided into two or more classes, and the learner must produce a model that assigns unseen inputs to one or more (multi-label classification) of these classes. This is typically tackled in a supervised manner. Spam filtering is an example of classification, where the inputs are email (or other) messages and the classes are "spam" and "not spam".
- In regression, also a supervised problem, the outputs are continuous rather than discrete.
- In clustering, a set of inputs is to be divided into groups. Unlike in classification, the groups are not known beforehand, making this typically an unsupervised task.

Density estimation finds the distribution of inputs in some space.

Dimensionality reduction simplifies inputs by mapping them into a lower-dimensional space. Topic modeling is a related problem, where a program is given a list of human language documents and is tasked with finding out which documents cover similar topics.

Among other categories of machine learning problems, learning to learn its own inductive bias based on previous experience. Developmental learning, elaborated for robot learning, generates its own sequences (also called curriculum) of learning situations to cumulatively acquire repertoires of novel skills through autonomous self-exploration and social interaction with human teachers and using guidance mechanisms such as active learning, maturation, motor synergies, and imitation.

2.4.3　The theory of Machine Learning

A core objective of a learner is to generalize from its experience. Generalization in this context is the ability of a learning machine to perform accurately on new, unseen examples/tasks after having experienced a learning data set. The training examples come from some generally unknown probability distribution (considered representative of the space of occurrences) and the learner has to build a general model about this space that enables it to produce sufficiently accurate predictions in new cases.

The computational analysis of machine learning algorithms and their performance is a branch of theoretical computer science known as computational learning theory. Because training sets are finite and the future is uncertain, learning theory usually does not yield guarantees of the performance of algorithms. Instead, probabilistic bounds on the performance are quite common. The bias–variance decomposition is one way to quantify generalization error.

For the best performance in the context of generalization, the complexity of the hypothesis should match the complexity of the function underlying the data. If the hypothesis is less complex than the function, then the model has underfit the data. If the complexity of the model is increased in response, then the training error decreases. But if the hypothesis is too complex, then the model is subject to overfitting and generalization will be poorer.

In addition to performance bounds, computational learning theorists study the time complexity and feasibility of learning. In computational learning theory, a computation is considered feasible if it can be done in polynomial time. There are two kinds of time complexity results. Positive results show that a certain class of functions can be learned in polynomial time. Negative results show that certain classes cannot be learned in polynomial time.

2.4.4　The approaches of Machine Learning

The following is a list of machine learning algorithms:
(1) Decision tree learning
Decision tree learning uses a decision tree as a predictive model, which maps

observations about an item to conclusions about the item's target value (shown in Figure 2.2).

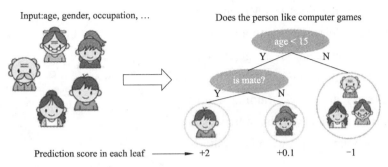

Decision tree learning

Figure 2.2　Sample of decision tree learning

(2) Association rule learning

Association rule learning is a method for discovering interesting relations between variables in large databases.

(3) Artificial neural networks

An artificial neural network (ANN) learning algorithm, usually called "neural network" (NN), is a learning algorithm that is vaguely inspired by biological neural networks. Computations are structured in terms of an interconnected group of artificial neurons, processing information using a connectionist approach to computation. Modern neural networks are non-linear statistical data modeling tools. They are usually used to model complex relationships between inputs and outputs, to find patterns in data, or to capture the statistical structure in an unknown joint probability distribution between observed variables.

Association rule learning

(4) Deep learning

Falling hardware prices and the development of GPUs for personal use in the last few years have contributed to the development of the concept of deep learning which consists of multiple hidden layers in an artificial neural network. This approach tries to model the way the human brain processes light and sound into vision and hearing. Some successful applications of deep learning are computer vision and speech recognition.

Deep learning

(5) Inductive logic programming

Inductive logic programming (ILP) is an approach to rule learning using logic programming as a uniform representation for input examples, background knowledge, and hypotheses. Given an encoding of the known background knowledge and a set of examples represented as a logical database of facts, an ILP system will derive a hypothesized logic program that entails all positive and no negative examples. Inductive programming is a related field that considers any kind of programming languages for representing hypotheses (and not only logic programming), such as functional programs.

Inductive logic programming

Support vector machines

(6) Support vector machines

Support vector machines (SVMs) are a set of related supervised learning methods used for classification and regression. It is given a set of training examples, each marked as belonging to one of two categories, an SVM training algorithm builds a model that predicts whether a new example falls into one category or the other (shown in Figure 2.3).

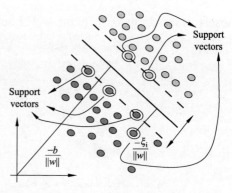

Figure 2.3　Sample of SVM

(7) Clustering

Cluster analysis is the assignment of a set of observations into subsets (called clusters) so that observations within the same cluster are similar according to some predesignated criterion or criteria, while observations drawn from different clusters are dissimilar. Different clustering techniques make different assumptions on the structure of the data, often defined by some similarity metric and evaluated for example by internal compactness (similarity between members of the same cluster) and separation between different clusters. Other methods are based on estimated density and graph connectivity. Clustering is a method of unsupervised learning, and a common technique for statistical data analysis.

(8) Bayesian networks

A Bayesian network, belief network or directed acyclic graphical model is a probabilistic graphical model that represents a set of random variables and their conditional independencies via a directed acyclic graph (DAG). For example, a Bayesian network could represent the probabilistic relationships between diseases and symptoms. Given symptoms, the network can be used to compute the probabilities of the presence of various diseases. Efficient algorithms exist that perform inference and learning.

(9) Reinforcement learning

Reinforcement learning is concerned with how an agent ought to take actions in an environment so as to maximize some notion of long-term reward. Reinforcement learning algorithms attempt to find a policy that maps states of the world to the actions the agent ought to take in those states. Reinforcement learning differs from the supervised learning problem in that correct input/output pairs are never presented, nor sub-optimal actions explicitly corrected.

(10) Representation learning

Several learning algorithms, mostly unsupervised learning algorithms, aim at discovering better representations of the inputs provided during training. Classical examples include principal components analysis and cluster analysis. Representation learning algorithms often attempt to preserve the information in their input but transform it in a way that makes it useful, often as a pre-processing step before performing classification or predictions, allowing reconstruction of the inputs coming from the unknown data generating distribution, while not being necessarily faithful for configurations that are implausible under that distribution.

Representation learning

Manifold learning algorithms attempt to do so under the constraint that the learned representation is low-dimensional. Sparse coding algorithms attempt to do so under the constraint that the learned representation is sparse (has many zeros). Multilinear subspace learning algorithms aim to learn low-dimensional representations directly from tensor representations for multidimensional data, without reshaping them into (high-dimensional) vectors. Deep learning algorithms discover multiple levels of representation, or a hierarchy of features, with higher-level, more abstract features defined in terms of (or generating) lower-level features. It has been argued that an intelligent machine is one that learns a representation that disentangles the underlying factors of variation that explain the observed data.

(11) Similarity and metric learning

In this problem, the learning machine is given pairs of examples that are considered similar and pairs of less similar objects. It then needs to learn a similarity function (or a distance metric function) that can predict if new objects are similar. It is sometimes used in Recommendation systems.

Similarities metrics in edureka

(12) Sparse dictionary learning

In this method, a datum is represented as a linear combination of basis functions, and the coefficients are assumed to be sparse. Let x be a d-dimensional datum, D be a d by n matrix, where each column of D represents a basis function. r is the coefficient to represent x using D. Mathematically, sparse dictionary learning means solving $x \approx D \times r$ where r is sparse. Generally speaking, n is assumed to be larger than d to allow the freedom for a sparse representation.

Learning a dictionary along with sparse representations is strongly NP-hard and also difficult to solve approximately. A popular heuristic method for sparse dictionary learning is K-SVD.

Sparse dictionary learning has been applied in several contexts. In classification, the problem is to determine which classes a previously unseen datum belongs to. Suppose a dictionary for each class has already been built. Then a new datum is associated with the class such that it's best sparsely represented by the corresponding dictionary. Sparse dictionary learning has also been applied in image de-noising. The key idea is that a clean image patch can be sparsely represented by an image dictionary, but the noise cannot.

Genetic algorithm

A rule-based systems

Learning classifier systems in a nutshell

(13) Genetic algorithms

A genetic algorithm (GA) is a search heuristic that mimics the process of natural selection, and uses methods such as mutation and crossover to generate new genotype in the hope of finding good solutions to a given problem. In machine learning, genetic algorithms found some uses in the 1980s and 1990s. Conversely, machine learning techniques have been used to improve the performance of genetic and evolutionary algorithms.

(14) Rule-based machine learning

Rule-based machine learning is a general term for any machine learning method that identifies, learns, or evolves "rules" to store, manipulate or apply, knowledge. The defining characteristic of a rule-based machine learner is the identification and utilization of a set of relational rules that collectively represent the knowledge captured by the system. This is in contrast to other machine learners that commonly identify a singular model that can be universally applied to any instance in order to make a prediction. Rule-based machine learning approaches include learning classifier systems, association rule learning, and artificial immune systems.

(15) Learning classifier systems

Learning classifier systems (LCS) are a family of rule-based machine learning algorithms that combine a discovery component (e.g. typically a genetic algorithm) with a learning component (performing either supervised learning, reinforcement learning, or unsupervised learning). They seek to identify a set of context-dependent rules that collectively store and apply knowledge in a piecewise manner in order to make predictions.

(16) Feature selection approach

Feature selection is the process of selecting an optimal subset of relevant features for use in model construction. It is assumed the data contains some features that are either redundant or irrelevant, and can thus be removed to reduce calculation cost without incurring much loss of information. Common optimality criteria include accuracy, similarity and information measures.

Words and Expressions

artificial intelligence [ˌɑːtɪˈfɪʃəl ɪnˈtelɪdʒəns]	人工智能
pattern recognition [ˈpætən ˌrekəɡˈnɪʃən]	模式识别
malicious [məˈlɪʃəs] spiteful and deliberately harmful	*adj.* 怀有恶意的；蓄意的；预谋的
optical character recognition (OCR) [ˈɒptɪkl ˈkærɪktə ˌrekəɡˈnɪʃən]	光学字符识别

conflate [kənˈfleɪt] to bring (things) together and fuse (them) into a single entity	*v.* 合并
data mining [deɪtə maɪnɪŋ]	数据挖掘
unsupervised learning [ˌʌnˈsjuːpəvaɪzd ˈlɜːnɪŋ]	非监督式学习
supervised learning [sjuːpəvaɪzd ˈlɜːnɪŋ]	监督式学习
semi-supervised learning [ˈsemi sjuːpəvaɪzd ˈlɜːnɪŋ]	半监督式学习
active learning [ˈæktɪv ˈlɜːnɪŋ]	主动学习
reinforcement learning [ˌriːɪnˈfɔːsmənt ˈlɜːnɪŋ]	强化学习
spam [spæm] unsolicited bulk electronic messages	*n.* 垃圾邮件
curriculum [kəˈrɪkjələm] the set of courses, coursework, and their content, offered at a school or university	*n.* 课程
repertoire [ˈrepətwɑː(r)] the set of skills, attributes, experiences, etc., possessed by a person	*n.*（计算机的）指令表
motor synergy [ˈməʊtə ˈsɪnədʒi]	运动协调
probability [ˌprɒbəˈbɪləti] (mathematics) a number, between 0 and 1, expressing the precise likelihood of an event happening	*n.* 概率
bias-variance decomposition [ˈbaɪəs ˈveəriəns ˈdiːkɒmpəˈzɪʃən]	偏置-方差分解
polynomial [ˌpɒlɪˈnəʊmiəl] able to be described or limited by a polynomial	*adj.* 多项式的
decision tree learning [dɪˈsɪʒən triː ˈlɜːnɪŋ]	决策树
association rule learning [əˌsəʊʃiˈeɪʃən ruːl ˈlɜːnɪŋ]	关联性规则学习
artificial neural network [ˌɑːtɪˈfɪʃəl ˈnjʊərəl ˈnetwɜːk]	人工神经网络
deep learning [diːp ˈlɜːnɪŋ]	深度学习
inductive logic programming [ɪnˈdʌktɪv ˈlɒdʒɪk ˈprəʊɡræmɪŋ]	归纳逻辑编程
support vector machine (SVM) [səˈpɔːt ˈvektə məˈʃiːn]	支持向量机
cluster analysis [ˈklʌstə əˈnæləsɪs]	聚类分析
Bayesian network [ˈbeɪziən ˈnetwɜːk]	贝叶斯网络

representation learning [ˌreprɪzen'teɪʃən 'lɜːnɪŋ]	表示学习
similarity and metric learning [ˌsɪmɪ'lærɪti ænd 'metrɪk 'lɜːnɪŋ]	相似性度量学习
sparse dictionary learning [spɑːs 'dɪkʃənəri 'lɜːnɪŋ]	稀疏字典学习
genetic algorithm [dʒɪ'netɪk 'ælɡərɪðəm]	遗传算法
rule-based machine learning [ruːl 'beɪst mə'ʃiːn 'lɜːnɪŋ]	基于规则的机器学习
learning classifier system ['lɜːnɪŋ 'klæsɪfaɪə 'sɪstəm]	学习分类器系统
feature selection approach ['fiːtʃə sɪ'lekʃən ə'prəʊtʃ]	特征选择方法

Questions for Discussion

1. What is machine learning? Which aspect of machine learning is more interested to you?

2. How many potential applications does machine learning have?

3. What algorithm of machine learning did you learn before?

Chapter 3 Mechanical Manufacturing

3.1 CAD/CAM/CAPP

3.1.1 Computer-aided Design (CAD)

Computer-aided design (CAD) can be defined as the use of computer systems to assist in the creation, modification, analysis, or optimization of a design. The computer systems consist of the hardware and software to perform the specialized design functions required by the particular user firm. The CAD hardware typically includes the computer, one or more graphics display terminals, keyboards, and other peripheral equipment. The CAD software consists of the computer programs to implement computer graphics on the system plus application programs to facilitate the engineering functions of the user company. Examples of these application programs include stress and strain analysis of components, dynamic response of mechanisms, heat-transfer calculations, and numerical control part programming. The collection of application programs will vary from one user firm to the next because their product lines, manufacturing processes and customer markets are different. These factors give rise to differences in CAD system requirements.

Computer-aided design

(1) Computer Applications in Design and Graphics

When using a CAD system, the designer can conceptualize the object to be designed more easily on the computer screen and can consider alternative designs or modify a particular design quickly to meet the necessary design requirements. The designer can then subject the design to a variety of engineering analyses and can identify potential problems (such as an excessive load or deflection). The speed and accuracy of such analyses far surpass what is available from traditional methods. The CAD user usually inputs data by keyboard and mouse to produce graphic images [1] on the computer screen that can be reproduced as paper copies [2] with a plotter or printer. When something is drawn once, it never has to be drawn again. It can be retrieved from a library, and can be duplicated, stretched, sized and changed in many ways without having to be redrawn. Cut and paste techniques are used as labor-saving aids.

In geometric modeling, a physical object or any of its parts is described mathematically or analytically. The designer first constructs a geometric model by giving commands that create or modify lines, surfaces, solids, dimension and text. Together, these present an accurate and complete two- or three-dimensional representation of the object. The results of these commands are displayed and can be moved around on the screen, and any section desired can be magnified to view details. These data are stored in the database contained in computer memory.

Engineers generally agree that the computer does not change the nature of the design process but is a significant tool that improves efficiency and productivity. The designer and the CAD system may be described as a design team: the designer provides knowledge creativity, and control; the computer generates accurate, easily modifiable graphics, performs complex design analysis at great speed, and stores and recalls design information. Occasionally, the computer may augment or replace many of the engineer's other tools, but it cannot replace the design process, which is controlled by the designer.

(2) Advantages of CAD

Depending on the nature of the problem and the sophistication of the computer system, CAD offer the designer or drafter some or all of the following advantages:

- Easier creation and correction of drawings. Engineering drawings may be created more quickly than by hand, and making changes and modifications is more efficient than correcting drawings made by hand.
- Better visualization of drawings. Many systems allow different views of the same object to be displayed and 3D pictorials [3] to be rotated on the computer screen.
- Database of drawing aids. Creation and maintenance of design databases (libraries of designs) permits storing designs and symbols for easy recall and application to the solution of new problems.
- Quick and convenient design analysis. Because the computer offers ease of analysis, the designer can evaluate alternative designs, thereby considering more possibilities while speeding up the process at the same time.

- Simulation and testing of designs. Some computer systems make the simulation of a product's operation possible, testing the design under a variety of conditions and stresses. Computer testing may improve on or replace models and prototypes.
- Increased accuracy. Computer is capable of producing drawings with more accuracy than by hand. Many CAD systems are even capable of detecting errors and informing the user of them.
- Improved filing. Drawings can be more conveniently filed, retrieved, and transmitted on disks and tapes.

Computer graphics has an almost limitless number of applications in engineering and other technical fields. Most graphical solutions that are possible with a pencil can be done on a computer and usually more productively.

3.1.2 Computer-aided Manufacturing (CAM)

Computer-aided manufacturing (CAM) can be defined as the use of computer systems to plan, manage, and control the operations of a manufacturing plant through either direct or indirect computer interface with the plant's production resources. As indicated by the definition, the applications of computer-aided manufacturing fall into two broad categories:

Computer-aided manufacturing

- Computer monitoring and control. These are the direct applications in which the computer is connected directly to the manufacturing process for the purpose of monitoring or controlling the process.
- Manufacturing support applications. These are the indirect applications in which the computer is used in support of the production operations in the plant, but there is no direct interface between the computer and the manufacturing process.

The distinction between the two categories is fundamental to an understanding of computer-aided manufacturing. It seems appropriate to elaborate on the brief definitions of the two types.

Computer monitoring and control can be separated into monitoring applications and control applications. Computer process monitoring involves a direct computer interface with the manufacturing process for the purpose of observing the process and associated equipment and collecting data from the process. The computer is not used to control the operation directly. The control of the process remains in the hands of human operators who may be guided by the information compiled by the computer.

Computer process control goes one step further than monitoring by not only observing the process but also controlling it based on the observations. With computer monitoring the flow of data between the process and the computer is in one direction only, from the process to the computer. In control, the computer interface allows for a two-way flow of data. Signals are transmitted from the process to the computer, just as in the of computer monitoring. In addition, the computer issues command signals directly to the manufacturing process based on control algorithms contained in its software.

(1) CAM hierarchy

A large scale CAM system contains a hierarchical structure of two or three levels of computers that are used to control and monitor individual process tasks. A small (mini-) computer is responsible for the management of a single processed information. This general configuration of a CAM hierarchy is illustrated in Figure 3.1.

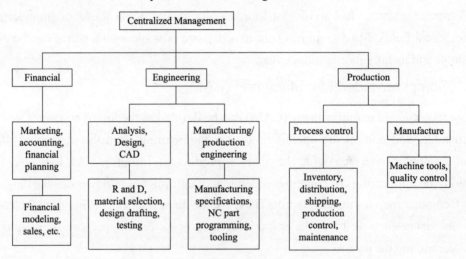

Figure 3.1 Computer aided manufacturing (CAM) hierarchical stucture

A large scale CAM system encompasses three major areas related to the manufacturing process: production management and control; engineering analysis and design; and finance and marketing [4]. Each is comprised of subtasks that are controlled either directly from a large computer (i.e. inventory control), or by a small computer, as in the case of inspection/quality control. Regardless of the control method, the important strength of CAM is that a two-way flow of information occurs.

Because the CAM system oversees many aspects of the manufacturing process, changes dictated by information monitored from one subtask can be translated into control data for some other subtask [5]. For example, in the manufacturing task, machining, inspection, and assembly are all under computer control. When the computer recognizes that a component is continually out of tolerance (based on information feedback from automated testing equipment), it can be programmed to effect a change in the actual machining process to compensate for the error. Since both subtasks become part of the same information loop, a feedback control system including machining and inspection is established.

(2) NC in computer-aided manufacturing

Although numerically controlled machine tools are essential for the development of operational CAM systems, those described in this text cannot be used in a computer based manufacturing system. Conventional NC (Numerical Control) and AC (Adaptive Control) machines must be modified so that information may be passed between the MCU and a computer based system. This modification has result in three major developments derived from the NC concept: computer managed numerical control, the cluster concept, and new forms of adaptive control.

Computer managed numerical control is a generic term that encompasses DNC (Direct Numerical Control) and CNC (Computerized Numerical Control). DNC and CNC are methods that distribute programmable computing responsibility between a control computer and NC machine tool. Neither system changes the functional characteristics of the NC machine; instead, each provides a means for communication of process data and commands outside the NC machine control loop [6].

The cluster concept is essentially an extension of computer management to more than one kind of machine [7]. A series of machine tools (e.g. those used for milling, boring grinding) are interconnected by a conveying system that automatically supplies individual machines with components at the required time. Two levels of control and monitoring become necessary. The individual machines are controlled with computer managed NC, and the cluster itself is managed by a centralized computer coordinating the production output of many clusters. Again, a hierarchical arrangement becomes evident.

Boring

Adaptive NC systems are part of the CAM environment. Process information is made available to centralized computer so that exceptional conditions (i.e. tool breakage) may be detected and corrected. In addition, adaptive feedback may also be recorded and analyzed so that the production efficiency of a given operation may be established.

Milling

3.1.3 Computer-aided Process Planning (CAPP)

Process planning represents the link between engineering design and shop-floor manufacturing. It is concerned with selecting methods of production: tooling, fixtures, machinery, sequence of operations, and assembly. All of these diverse activities must be planned, which traditionally has been done by process planners. The sequence of processes and operations to be performed, the machines to be used, the standard time for each operation and similar information are all documented on a routing sheet.

Computer-aided process planning

When done manually, this task is highly labor-intensive and time-consuming and relies heavily on the experience of the process planner. Modern practice in routing sheets is to store the relevant data in computers and affix a bar code (or other identification) to the part. In order to shorten the gap between CAD and CAM, many computerized process planning approaches have been developed to accomplish the task of process planning in the past decades. These approaches are known as computer-aided process planning (CAPP). With the help of CAPP, better and faster process plans can be generated. CAPP can be classified into three categories: indexed CAPP, variant CAPP and generative CAPP. Each approach is appropriate under certain conditions [8].

(1) Advantages of CAPP

A planner must manage and retrieve a great deal of data and many documents, including established standards, machinability data, machine specifications, tooling inventories, stock availability, and existing process plans. This is primarily an information-handling job, and the computer is an ideal companion.

There is another advantage to using computers to help with process planning. Because the task involves many interrelated activities, determining the optimum plan requires many iterations. Since computers can readily perform vast numbers of comparisons, many more alternative plans can be explored than would be possible manually. A third advantage in the use of computer-aided process planning is uniformity.

Several specific benefits can be expected from the adoption of compute-aided process planning techniques as follows:

- Reduced clerical effort in preparation of instructions.
- Fewer calculation errors due to human error.
- Fewer oversights in logic or instructions because of the prompting capability available with interactive computer programs.
- Immediate access to up-to-date information from a central database.
- Consistent information, because every planner accesses the same database.
- Faster response to changes requested by engineers of other operating departments. Automatic use of the latest revision of a part drawing.
- More-detailed, more-uniform process-plan statements produced by word-processing techniques.
- More-effective use of inventories of tools, gages, and fixtures and a concomitant reduction in the variety of those items.
- Better communication with shop personnel because plans can be more specifically tailored to a particular task and presented in unambiguous, proven language.
- Better information for production planning, including cutter life, forecasting materials requirements planning, scheduling, and inventory control.

(2) Approaches to computer-aided process planning

Most important for Computer integrated manufacturing (CIM), computer-aided process planning produces machine readable data instead of handwritten plans. Such data can readily be transferred to other systems within the CIM hierarchy for use in planning.

There are basically two approaches to computer-aided process planning: variant and generative.

In the variant approach, a set of standard process plans is established for all the parts families that have been identified through group technology (GT). The standard plans are stored in computer memory and retrieved for new parts according to their family identification. Again, GT helps to place the new part in an appropriate family. The standard plan is then edited to suit the specific requirements of a particular Job.

In the generative approach, an attempt is made to synthesize each individual plan using appropriate algorithms that define the various technological decisions that must be made in the course of manufacturing. In a truly generative process planning system, the sequence of operations, as well as all the manufacturing process parameters, would be automatically established without reference to prior plans. In its ultimate realization, such an approach would be universally applicable: present any plan to the system, and the computer produces the optimum process plan.

No such system exists, however. So called generative process-planning systems (and probably for the foreseeable future) are still specialized systems developed for a specific operation or a particular type of manufacturing process. The logic is based on a combination of past practice and basic technology.

Words and Expressions

alternative design [ɔːlˈtɜːnətɪv dɪˈzaɪn]	替代设计（方案）；可供选择的设计（方案）
augment [ɔːgˈment] enlarge or increase	v. 增加；提高
clerical [ˈklerɪkl] appropriate for or engaged in office work	adj. 文书的；办公室工作的
cluster [ˈklʌstə(r)] a grouping of a number of similar things	n. 群；组
companion [kəmˈpænjən] one is frequently in the company of another	n. 同伴
computer-aided process planning (CAPP) [kəmˈpjuːtə eɪdɪd ˈprəʊses ˈplænɪŋ]	计算机辅助工艺规划
conceptualize [kənˈseptʃuəlaɪz] to form an idea	n. 概念化；抽象化

concomitant [kənˈkɒmɪtənt] following as a consequence	*adj.* 伴随的；同时发生或出现的
dictate [dɪkˈteɪt] issue commands or orders for sb.	*v.* 命令
filing [ˈfaɪlɪŋ] the entering of a legal document into the public record	*n.* 存档档案；归档；存档
generative [ˈdʒenərətɪv] having the ability to produce or originate	*adj.* 再生的；创成的
hierarchy [ˈhaɪərɑːki] a series of ordered groupings of people or things within a system	*n.* 体系；层次；分级结构
inventory [ˈɪnvəntri] a detailed list of all the items in stock	*n.* 库存；存货清单
iteration [ˌɪtəˈreɪʃən] (computer science) a single execution of a set of instructions that are to be repeated	*n.* 迭代；逐步逼近法
limitless [ˈlɪmɪtləs] without limits in extent or size or quantity	*adj* 无限的；无界限的
MCU (monitor and control unit) [ˈmɒnɪtə ænd kənˈtrəʊl ˈjuːnɪt]	监控设备
optimization [ˌɒptɪmaɪˈzeɪʃən] the act of rendering optimal	*n.* 优化
oversee [ˌəʊvəˈsiː] watch and direct	*v.* 监督；监察
oversight [ˈəʊvəsaɪt] an unintentional omission resulting from failure to notice something	*n.* 疏忽；失察；监管
parameter [pəˈræmɪtə(r)] a constant in the equation of a curve that can be varied to yield a family of similar curves	*n.* 系数；特征值
pictorial [pɪkˈtɔːriəl] using or containing pictures	*adj.* 图示的
plotter [ˈplɒtə] a device that draws up a graph or chart	*n.* 绘图仪；绘图机

prompt [prɒmpt] give a cue a performer (usually the beginning of the next line to be spoken)	v. 提醒；提示
simulation [ˌsɪmjʊˈleɪʃən] (computer science) the technique of representing the real world by a computer program	n. 仿真；模拟
standard process plan [ˈstændəd ˈprəʊses ˈplæn]	典型工艺规程
tailor [ˈteɪlə(r)] to make fit for a specific purpose	v. 定做；满足……（的要求，需要，条件等）
tolerance [ˈtɒlərəns] the power or capacity of an organism to tolerate unfavorable environmental conditions	n. 公差
unambiguous [ˌʌnæmˈbɪɡjuəs] having or exhibiting a single clearly defined meaning	adj. 明确的；清楚的；无歧义的
up-to-date [ˈʌp tuː deɪt] reflecting the latest information or changes	adj. 现代的；最新的；当今的
variant [ˈveərɪənt] an event that departs from expectations	n. 变形；变体

Notes

[1] graphic image 可译为"图形、图像"。

[2] paper copy 可译为"打印输出"。

[3] 3D pictorial 可译为"三维图像"。

[4] 本句可译为：一个大规模的 CAM 系统包含三个与制造工艺过程有关的主要领域：生产管理与控制，工程分析与设计，财务与市场。

[5] 本句可译为：因为 CAM 系统管理制造工艺过程的很多方面，所以由一个子任务监测到的信息所支配的变化，能够转换为其他子任务的控制数据。

[6] 本句可译为：两个系统都不改变数控机床的功能特性，而是每一系统都提供了在数控机床控制回路之外的工艺过程数据与命令的通信方法。

[7] 本句可译为：群集概念实质上是对多种机床的计算机管理的延伸。

[8] 这几句可译为：为了将 CAD 和 CAM 连成一体，过去几十年里已开发出许多采用计算机来进行工艺规划的手段，它们被称为计算机辅助工艺规划（CAPP）。借助于 CAPP，工艺规划可以做得更快更好。CAPP 可以分为检索式、样件式和创成式三种类型，它们都有各自的适用条件。

Questions for Discussion

1. What are the advantages of CAD/CAM/CAPP?
2. What is the relationship between NC and CAM?
3. Why can better and faster process plans be generated with the help of CAPP?

3.2 Flexible Manufacturing System

3.2.1 The Definition of Flexible Manufacturing System

Flexible manufacturing system

The evolution of manufacturing can be represented graphically as a continuum as shown in Figure 3.2. The manufacturing processes and systems are in a state of transition from manual operation to the eventual realization of fully integrated manufacturing. Flexible manufacturing is the step between islands of automation and computer integrated manufacturing.

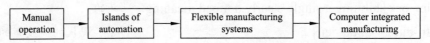

Figure 3.2 The evolution of manufacturing

Flexibility is an important characteristic in the modern manufacturing setting. It means that a manufacturing system is versatile and adaptable, while also capable of handling relatively high production runs. A flexible manufacturing system is versatile in that it can produce a variety of parts. It is adaptable in that it can be quickly modified to produce a completely different line of parts. This flexibility can be the difference between success and failure in a competitive international marketplace.

It is a matter of balance. Stand-alone computer numerical control (CNC) machines have a high degree of flexibility, but are capable of relatively low-volume production runs. At the opposite end of the spectrum, transfer lines are capable of high-volume runs, but they are not very flexible. Flexible manufacturing is an attempt to use technology in such a way as to achieve the optimum balance between flexibility and production runs[1]. These technologies include automated materials handling, group technology, and computer and distributed numerical control.

A flexible manufacturing system (FMS) is an individual machine or group of machines served by an automated materials handling system that is computer controlled and has a tool handling capability [2]. Because of its tool handling capability and computer control, such a system can be continually reconfigured to manufacture a wide variety of parts[3]. This is why it is called a flexible manufacturing system.

The key elements necessary for a manufacturing system to qualify as an FMS are as

follows:
- Computer control;
- Automated materials handling capability;
- Tool handling capability.

Flexible manufacturing represents a major step toward the goal of fully integrated manufacturing in that it involves integration of automated production processes. In flexible manufacturing, the automated manufacturing machine (i. e. lathe, mill, drill) and the automated materials handling system share instantaneous communication via a computer network. This is integration on a small scale.

3.2.2 The Overview of Flexible Manufacturing

Flexible manufacturing takes a major step toward the goal of fully integrated manufacturing by integrating several automated manufacturing concepts as follows:
- Computer numerical control (CNC) of individual machine tools;
- Distributed numerical control (DNC) of manufacturing systems;
- Automated materials handling systems;
- A Group technology (families of parts).

When these automated processes, machines, and concepts are brought together in one integrated system, an FMS is the result. Humans and computers play major roles in an FMS. The amount of human labor is much less than with a manually operated manufacturing system, of course. However, humans still play a vital role in the operation of an FMS. Human tasks include the followings:
- Equipment troubleshooting, maintenance, and repair;
- Tool changing and setup;
- Loading and unloading the system;
- Data input;
- Changing of parts programs;
- Development of programs.

Flexible manufacturing system equipment, like all manufacturing equipment, must be monitored for "bugs", malfunctions, and breakdowns. When a problem is discovered, a human troubleshooter must identify its source and prescribe corrective measures. Humans also undertake the prescribed measures or repair the malfunctioning equipment. Even when all systems are properly functioning, periodic maintenance is necessary.

Human operators also set up machines, change tools, and reconfigure systems as necessary. The tool handling capability of an FMS decreases, but does not eliminate human involvement in tool changing and setup. The same is true of loading and unloading the FMS. Once raw material has been loaded onto the automated materials handling system, it is moved through the system in the prescribed manner. However, the original loading onto the materials handling system is still usually done by human operators, as is the unloading of finished products.

Humans are also needed for interaction with the computer. Humans develop parts programs that control the FMS via computers. They also change the programs as necessary when reconfiguring the FMS to produce another type of part or parts: humans also input data needed by the FMS during manufacturing operations. Humans play less labor-intensive an FMS, but the roles are still critical.

Control at all levels in an FMS is provided by computers. Individual machine tools within an FMS are controlled by CNC. The overall system is controlled by DNC. The automated materials handling system is computer controlled, as are other function including data collection, system monitoring, tool control, and traffic control. Human-computer interaction is the key to the flexibility of an FMS.

3.2.3 The Rationale for Flexible Manufacturing

In manufacturing there have always been trade-offs between production rates and flexibility. At one end of the spectrum are transfer lines capable of high production rates, but low flexibility. At the other end of the spectrum are independent CNC machines that offer maximum flexibility, but are only capable of low production rates. Note from the figure that flexible manufacturing falls in the middle of the continuum. There has always been a need in manufacturing for a system that could produce higher volume and production runs than could independent machines while still maintaining flexibility.

Transfer lines are capable of producing large volumes of parts at high production rates. The line takes a great deal of setup, but can turn out identical parts in large quantities. Its chief shortcoming is that even minor design changes in a part can cause the entire line to be shut down and reconfigured. This is a critical weakness because it means that transfer lines cannot produce different parts, even parts from within the same family, without costly and time-consuming shutdown and reconfiguration.

Traditionally, CNC machines have been used to produce small volumes of parts that differ slightly in design. Such machines are ideal for this purpose because they can be quickly reprogrammed to accommodate minor or even major design changes. However, as independent machines, they cannot produce parts in large volumes or at high production rates.

An FMS can handle higher volumes and production rates than independent CNC machines. They cannot quite match such machines for flexibility, but they come close. What is particularly significant about the middle ground capabilities of flexible manufacturing is that most manufacturing situations require medium production rates to produce medium volumes with enough flexibility to quickly reconfigure to produce another part or product. Flexible manufacturing fills this longstanding void in manufacturing.

Flexible manufacturing, with its middle ground capabilities, offers a number of advantages for manufacturers as follows:
- Flexibility within a family of parts;
- Random feeding of parts;
- Simultaneous production of different parts;
- Decreased setup times / lead time;
- Efficient machine usage;
- Decreased direct and indirect labor costs;
- Ability to handle different materials;
- Ability to continue some production if one machine breaks down.

3.2.4 The Components of Flexible Manufacturing System

(1) Machine tools

A flexible manufacturing system uses the same types of machine tools as any other manufacturing system, be it automated or manually operated. These include lathes, mills, drills, saws, and so on. The type of machine tools actually included in an FMS depends on the setting in which the system will be used. Some FMSs are designed to meet a specific, well-defined need. In these cases, the machine tools included in the system will be only those necessary for the planned operations. Such a system would be known as a dedicated system. In a job shop setting, or any other setting in which the actual application is not known ahead of time or must necessarily include a wide range of possibilities, machines capable of performing at least the standard manufacturing operations would be included. Such systems are known as general-purpose systems.

(2) Control systems

The control system for an FMS serves a number of different control functions for the system as follows:
- Storage and distribution of part programs;
- Work flow control and monitoring;
- Production control;
- System/tool control/ monitoring.

(3) Materials handling system

The automated materials handling system is a fundamental component that helps mold a group of independent CNC machines into a comprehensive FMS. The system must be capable of accepting workpieces mounted on pallets and moving them from workstation to workstation as needed. It must also be able to place workpieces "on hold" as they wait to be processed at a given workstation.

The materials handling system must be able to unload a workpiece at one station and load another for transport to the next station. It must accommodate computer control and be completely compatible in that regard with other components in the flexible manufacturing system. Finally, the materials handling system for an FMS must be able to withstand the rigors of a shop environment.

Some FMSs are configured with automated guided vehicles (AGVs) as a principal means of materials handling.

(4) Human operators

The final component in an FMS is the human component. Although flexible manufacturing as a concept decreases the amount of human involvement in manufacturing, it does not eliminate it completely. Further, the roles humans play in flexible manufacturing are critical roles.

Words and Expressions

accommodate [əˈkɒmədeɪt] be agreeable or acceptable to sth.	v. 适应
automated guided vehicle (AGV) [ˈɔːtəmeɪtɪd ˈɡaɪdɪd ˈviːɪkl]	自动导向小车
be compatible with [bi: kəmˈpætəbl wɪð]	相容；兼容；适合于
bug [bʌɡ] a fault in a machine	n. 缺陷；错误
continuum [kənˈtɪnjuəm] a continuous nonspatial whole or extent or succession in which no part or portion is distinct of distinguishable from adjacent parts	n. 连续（统一体）
dedicated [ˈdedɪkeɪtɪd] designed to do only one particular type of work	adj. 专用的
family of parts [ˈfæməli əv pɑːts]	相似部件；零件族（组）
flexible manufacturing [ˈfleksəbl ˌmænjʊˈfæktʃərɪŋ]	柔性制造
labor-intensive [ˈleɪbər ɪnˈtensɪv] an industry or type of work that needs a lot of workers	adj. 劳动强度大的；劳动密集的
lead time [liːd taɪm]	产品设计至实际投产间的时间；提前期；生产准备期

longstanding [ˌlɒŋˈstændɪŋ] having continued or existed for a long time	*adj.* 长久持续的；长期存在的
periodic [ˌpɪəriˈɒdɪk] happening or recurring at regular intervals	*adj.* 周期的；定期的；循环的
realization [ˌriːəlaɪˈzeɪʃən] coming to understand something clearly and distinctly	*n.* 实现
reconfigure [ˌriːkənˈfɪɡə(r)] to rearrange the elements or settings of (a system, device, computer application, etc.)	*v.* 重新配置（组合）
rigor [ˈrɪɡə] something hard to endure	*n.* 艰苦；严酷
simultaneous [ˌsɪməlˈteɪniəs] occurring or operating at the same time	*adj.* 同步的；同时发生的
stand-alone [stænd əˈləʊn] capable of operating independently	*adj.* 可独立应用的
time-consuming [taɪm kənˈsjuːmɪŋ] laking a long time and patience to do sth.	*adj.* 费时的；拖延时间的
trade-off [ˈtreɪd ɒf] an exchange that occurs as a compromise	*n.* 折中（办法，方案）；权衡
traditional [trəˈdɪʃənl] consisting of or derived from tradition	*adj.* 传统的；惯例的
transfer line [trænsˈfɜː laɪn]	组合机床自动线
transition [trænˈzɪʃən] the act of passing from one state or place to the next	*n.* 转变；转换
troubleshooter [ˈtrʌblʃuːtə(r)] a worker whose job is to locate and fix sources of trouble (especially in mechanical devices)	*n.* 故障检修工
troubleshooting [ˈtrʌblʃuːtɪŋ] a form of problem solving, often applied to repair failed products or processes	*n.* 发现并修理故障，排除故障
volume production [ˈvɒljuːm prəˈdʌkʃən]	批量生产
workpiece [ˈwɜːkˌpiːs] work consisting of a piece of metal being machined	*n.* 工件

Notes

[1] 本句可译为：柔性制造试图按这样一种方式来使用技术，以取得灵活性与流水生产之间的最佳平衡。

[2] 本句可译为：柔性制造系统是由计算机控制的自动物料搬运系统服务的单台机床或组机床，并拥有刀具装卸能力。

[3] 本句可译为：由于它的刀具装卸能力和计算机控制，所以能够不断重新配置这样的系统去制造种类繁多的零件。

Questions for Discussion

1. What is a flexible manufacturing system (FMS)?
2. What are the key elements that make a manufacturing system an FMS?
3. Which four concepts are integrated in an FMS?
4. List and explain at least four tasks humans accomplished in an FMS.

3.3 Computer Numerical Control

3.3.1 Introduction

In 1968, a CNC (computer numerical control) machine was marketed, which could automatically change tools, so that many different processes could be done on one machine. Such a machine became known as a machining center—a multifunctional CNC machine that can perform a variety of processes and change tools automatically while under programmable control. The computer-controlled machining centers have the required flexibility and versatility that other individual machine tools do not have, so they often become the first choice in machine-tool selection.

The CNC machines still perform essentially the same functions as manually operated machine centers, but movements of the machine tool are controlled electronically rather than by hand. CNC machine centers can produce the same parts over and over again with very little variation. They can run day and night, week after week, without getting tired. These are obvious advantages over manually operated machine centers, which need a great deal of human interaction in order to do anything.

A CNC machine tool differs from a manually operated machine tool only in respect to the specialized components that make up the CNC system. The CNC system can be further divided into three subsystems: control, drive, and feedback. All of these subsystems must work together to form a complete CNC system.

3.3.2 CNC System

(1) Control system

The centerpiece of the CNC system is the control. Technically the control is called the machine control unit (MCU), but the most common names used in recent years are controller control unit, or just plain control. This is the computer that stores and reads the program and tells the other components what to do.

(2) Drive system

The drive system is comprised of screws and motors that will finally turn the part program into motion [1]. The first component of the typical drive system is a high-precision lead screw called a ball screw. Eliminating backlash in a ball screw is very important for two reasons. First, high-precision positioning cannot be achieved if the table is free to move slightly when it is supposed to be stationary. Second, material can be climb-cut safely if the backlash has been eliminated. Climb cutting is usually the most desirable method for machining on a CNC machine tool.

Drive motors are the second specialized component in the drive system. The turning of the motor will turn the ball screw to directly cause the machining table to move [2]. Several types of electric motors are used on CNC control systems, and hydraulic motors are also occasionally used.

The simplest type of electric motor used in CNC positioning systems is the stepper motor (sometimes called a stepping motor). A stepper motor rotates a fixed number of degrees when it receives an electrical pulse and then stops until another pulse is received [3]. The stepping characteristic makes stepper motors easy to control.

It is more common to use servomotors in CNC systems today. Servomotors operate in a smooth, continuous motion—not like the discrete movements of the stepper motors. This smooth motion leads to highly desirable machining characteristics, but they are also difficult to control. Specialized hardware controls and feedback systems are needed to control and drive these motors. Alternating current (AC) servomotors are currently the standard choice for industrial CNC machine tools [4].

(3) Feedback system

The function of a feedback system is to provide the control with information about the status of the motion control system, which is described in Figure 3.3.

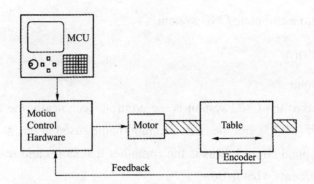

Figure 3.3　Typical motion control system of a CNC machine tool

The control can compare the desired condition to the actual condition and make corrections. The most obvious information to be fed back to the control on a CNC machine tool is the position of the table and the velocity of the motors. Other information may also be fed back that is not directly related to motion control, such as the temperature of the motor and the load on the spindle—this information protects the machine from damage.

There are two main types of control systems: open-loop and closed-loop [5]. An open-loop system does not have any device to determine if the instructions were carried out. For example, in an open-loop system, the control could give instructions to turn the motor 10 revolutions [6]. However, no information can come back to the control to tell it if it actually turned. All the control knows is that it delivered the instructions. Open-loop control is not used for critical systems, but it is a good choice for inexpensive motion control systems in which accuracy and reliability are not critical feedback uses external sensors to verify that certain conditions have been met. Of course, positioning and velocity feedback is of primary importance to an accurate CNC system. Feedback is the only way to ensure that the machine is behaving the way the control intended it to behave.

3.3.3　Types of CNC Machining Centers

There are various designs for machining centers. The two basic types are the vertical type and the horizontal type. Their names are derived from their respective spindle designs, which are in either a vertical position or horizontal position. There are other designs and variations of both vertical and horizontal machines that are designed for other application needs.

(1) Vertical-spindle machining centers

It is also called vertical machining centers (VMC). The VMC spindle holds the cutting tool in a vertical position. VMCs are generally used to perform operations on flat parts that require cutting on the top surface of the part and on parts with deep cavities, such as in mold and die making. A vertical-spindle machining center (which is similar to a vertical-spindle milling machine) is shown in Figure 3.4. The tool magazine is on the left of the figure, and all operations and movements are controlled and modified through the computer-control panel shown on the right. The VMC is programmed to position and cut in the X-, Y- and Z-axes (three-axes). Other options can be added that will increase the flexibility and productivity of the VMC. Some of the options that are typically added to VMCs include an indexer or a shuttle table. The indexer, which is mounted on the machine table can rotate the part along the X-axis for 360° machining (fourth axis). The shuttle table option is used for shuttling the parts, which allows the operator to load and unload parts without interrupting production. Because the thrust forces in vertical machining are directed downward, such machines have high stiffness and produce parts with good dimensional accuracy. These machines generally are less expensive than horizontal-spindle machines.

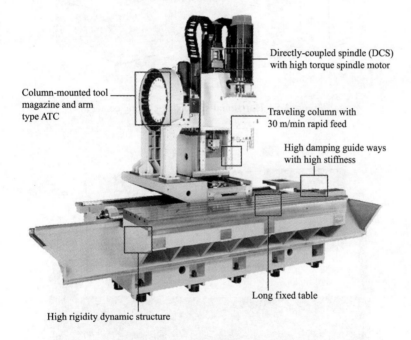

Vertical machining center

Figure 3.4 Vertical-spindle machining center

(2) Horizontal-spindle machining centers

It is also called horizontal machining centers (HMC). The HMC spindle holds the cutting tool in a horizontal position, which is suitable for large as well as tall workpieces that require machining on a number of their surfaces (shown in Figure 3.5). The HMC machine table can be programmed to rotate 360° in a circular motion (B-axis), so it is programmed to position and cut in the X-,Y-, Z- and B-axes (four-axes) The HMCs can machine parts on more than one side in one clamping, and find wide use in flexible manufacturing systems. The pallet can be swiveled on different axes to various angular positions. Another category of horizontal-spindle machines is turning centers, which are computer-controlled lathes with several features. A three-turret turning center is shown in Figure 3.6. It is constructed with two horizontal spindles and three turrets equipped with a variety of cutting tools used to perform several operations on a rotating workpiece.

Horizontal machining center

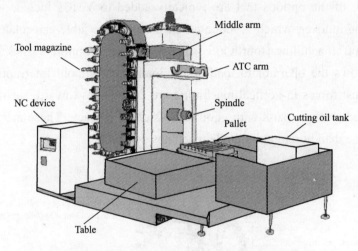

Figure 3.5　Horizontal-spindle machining center

Turning center

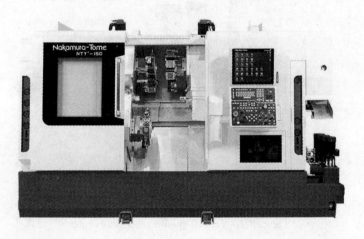

Figure 3.6　A three-turret turning center

(3) Universal machining centers

Universal machining centers are equipped with both vertical and horizontal spindles. They have a variety of features and are capable of machining all of the surfaces of a workpiece (that is, vertical and horizontal and at a wide range of angles).

3.3.4　Characteristics and Capabilities of CNC Machining Centers

The major characteristics of machining centers are summarized as follows:

- Machining centers are capable of handling a wide variety of part sizes and shapes efficiently, economically, repetitively, and with high dimensional accuracy—with tolerances in the order of ±0.0025 mm.
- These machines are versatile and capable of quick change-over from one type of product to another.
- The time required for loading and unloading workpieces, changing tools, gaging of the part, and troubleshooting is reduced. Therefor productivity is improved, thus reducing labor requirements (particularly skilled labor) and minimizing production costs.
- These machines are equipped with tool-condition monitoring devices for the detection of tool breakage and wear as well as probes for tool-wear compensation and tool positioning.
- In-process and post-process gaging and inspection of machined workpieces are now features of machining centers.
- These machines are relatively compact and highly automated and have advanced control systems, so one operator can attend to two or more machining centers at the same time thus reducing labor costs.

Machining centers are available in a wide variety of sizes and features. Typical spindle speed is 8,000 rpm (round per minute), and sometime can be as high as 75,000 rpm for special applications using small-diameter cutters. Modern spindle can accelerate to a speed of 20,000 rpm in only 1.5 seconds. Some pallets are capable of supporting workpieces weighing as much as 7,000 kg although even higher capacities are available for special applications.

Words and Expressions

alternating current [ˈɔːltəneɪtɪŋ ˈkʌrənt]	交流电
backlash [ˈbæklæʃ] a movement back from an impact	n. 反向间隙；回程误差
ball screw [bɔːl skruː]	滚珠丝杠
centerpiece [ˈsentəpiːs] the central or most important feature	n. 引人注目的东西；主要特征或最重要的特征
climb cut [klaɪm kʌt]	顺铣
die [daɪ] a device used for shaping metal	n. 模具
drive motor [draɪv ˈməʊtə]	驱动电动机
drive system [draɪv ˈsɪstəm]	传动系统；驱动系统
feedback [ˈfiːdbæk] response to an inquiry or experiment	n. 反馈

flexibility [ˌfleksəˈbɪləti] the quality of being adaptable or variable	*n.* 柔性；弹性
frame [freɪm] a structure supporting or containing something	*n.* 机架
gaging [ˈɡeɪdʒɪŋ] a measuring instrument for measuring and indicating a quantity such as the thickness of wire or the amount of rain etc.	*n.* 测量
horizontal machining center [ˌhɒrɪˈzɒntl məˈʃiːnɪŋ ˈsentə]	卧式加工中心
hydraulic motor [haɪˈdrɔːlɪk ˈməʊtə]	液压电机（作连续回转运动并输出转矩的液压执行元件）
indexer [ˈɪndeksə] sth. providing an index	*n.* 分度器
individual [ˌɪndɪˈvɪdʒuəl] considered separately rather than as part of a group	*adj.* 个别的；独特的
lead screw [liːd skruː]	丝杠
pallet [ˈpælət] a portable platform for storing or moving goods that are stacked on it	*n.* 托盘
servomotor [ˈsɜːvəʊˈməʊtə] a motor that supplies power to a servomechanism	*n.* 伺服电动机
stationary [ˈsteɪʃənəri] standing still	*adj.* 静止的，不动的
stepper motor [ˈstepə ˈməʊtə]	步进电机
subsystem [ˈsʌbˈsɪstəm] a system that is part of some larger system	*n.* 子系统
swivel [ˈswɪvl] a coupling (as in a chain) that has one end that turns on a headed pin	*v.* （使）旋转
tool magazine [tuːl ˌmæɡəˈziːn]	刀库
turning center [ˈtɜːnɪŋ ˈsentə]	车削中心
universal machining center [ˌjuːnɪˈvɜːsəl məˈʃiːnɪŋ ˈsentə]	万能加工中心

versatility [ˌvɜːsəˈtɪləti] having a wide variety of skills	*n.* （才能、用途等）多面性；通用性
vertical machining center [ˈvɜːtɪkl məˈʃiːnɪŋ ˈsentə]	立式加工中心

Notes

[1] screw 指"丝杠"，全句可译为：驱动系统由丝杠和电动机组成，它最终会把零件加工程序转化为运动。

[2] machining table 指"机床工作台"，全句可译为：电动机的转动会使滚珠丝杠转动，并使机床工作台产生运动。

[3] electrical pulse 指"电脉冲"，全句可译为：步进电机接收到一个电脉冲就会转动一个固定的角度，然后停止转动，直至接收到另一个脉冲。

[4] alternating current (AC) servomotor 指"交流伺服电动机"，全句可译为：目前，工业生产中使用的计算机数控机床上通常选用交流伺服电动机。

[5] open-loop and closed-loop 指"开环和闭环"，全句可译为：主要有两种类型的控制系统，即开环系统和闭环系统。

[6] revolution 指"转数"，全句可译为：例如，在开环系统中，控制装置可以发出让电动机转动 10 转的指令。

Questions for Discussion

1. When did people begin to develop the CNC machines and what communication medium did they often use?

2. According to the text, is CNC a kind of machine tool or a technique for controlling a wide variety of machines?

3. Why did people use feedback in the CNC systems?

3.4　Rail Grinding

3.4.1　Introduction

Rail grinding can prolong the service life of rails and improve the riding stability of a rail vehicle. Nowadays, the research on rail grinding is mainly focused on these three aspects: firstly, summing up and obtaining the influence of grinding on the rail based on long-term grinding experience in practice, which could provide reasonable guidance for the appropriate selection of a grinding; secondly, research on rail grinding regarding improvement of the rail service property, such as repair rail damage, profile optimization of rail grinding, and improvement of the wheel/rail contact performance of rail vehicles; thirdly, research on the rail grinding mechanism, such as thermodynamics during the rail grinding, factors that influence grinding quality, the influence of grinding wheel properties on the grinding result, and the rail forming grinding mechanism. At present, the research on rail grinding is mainly focused on the analysis of grinding mechanism and the factors that influence grinding quality based on the literature above. Furthermore, there are few studies on the practical grinding above studies cannot provide direct support for grinding operations at grinding sites and the improvement of rail grinding quality is limited to some extent.

In the practical rail grinding process, a corresponding target profile of rail grinding needs to be designed based on the damage and wear status of the on-site rail. Then a reasonable grinding mode that includes appropriate grinding parameters of a grinding train is selected based on experience to ensure that the obtained final grinding profile is consistent with the pre-designed target profile. However, selection of the grinding mode method based on experience cannot fully guarantee that the obtained final grinding profile will be consistent with the target profile. Therefore, it is necessary to predict the final grinding profile obtained with a certain grinding mode before practical grinding construction, which could provide a basis for the selection and optimization of the rail grinding mode.

3.4.2 Current Rail Grinding Practice

The main countries that study rail grinding technology are the USA, the UK, Germany, Switzerland, China and so on. Current network rail grinding operations include preventative and corrective re-profiling of the rail head. Preventative re-profiling is to re-profile the track and remove small defects using small depths of cut, typically less than 0.5 mm. This is performed on both straight track and curved sections. Corrective re-profiling is performed on sections of heavily worn / plastically deformed track or defective rail. Its primary purpose is to re-profile the track to its original form and it is usually performed on shorter sections of track with larger depths of cut. The different operations will be identified as preventative or corrective as follows.

Both these operations are currently done at varying rates with train speeds ranging from 1 to 10 mile/h. Opportunities exist to increase train speed, for both preventative and corrective operations, and depth of cut, where appropriate for corrective work, to improve productivity of these operations through the use of more advanced grinding technologies. Although the current depths of cut for preventative re-profiling are determined by the material removal required from the rail head and not the grinding process; an increased capability for larger cut depths per grinding stone would facilitate minimizing the stone requirements per train that could also lead to cost savings. As the largest potential benefit for

Network Rail exists in improving their preventative re-profiling operations, all modelling is for the preventative rail grinding process with the primary focus concerned with an increase in train speed for this operation.

New technologies have been introduced into the industry; they include milling and planing processes that show good potential for achieving the large cut depths that are particularly important in corrective re-profiling operations. These were assessed in the InnoTrack project and guidelines for their use were issued. However, the majority of UK re-profiling requirements are preventative operations on longer sections of track and they would greatly benefit from increased train speeds. Developments have been made here as well, with the introduction of the "high speed" grinding technique involving "non-driven" stones that removes a very shallow depth of cut. Although a significant step forward, this is better suited to networks with long sections of tangent track that are found in the primary user's country, Germany. The UK infrastructure is more complicated with a significant amount of curved sections, and varying cut depth requirements depending upon track usage [1]. Driven stones are still required to accommodate these different conditions and this work aims to investigate best practice techniques from the aerospace industry that best fits the UK's needs for these operations. The possibility of using peripheral grinding (i.e. using the outer edge of the stone) rather than face grinding has the potential to offer additional benefits and was also investigated.

3.4.3 Rail Grinding Mechanisms

In order to repair a worn rail profile by using the end surface of a grinding wheel, a grinding train has to place a group of high-speed spinning grinding wheels above the rail at different angles and to exert axial pressure to grind the profile by using the end surface of the grinding stone. The rail grinding mechanism is shown in Figure 3.7.

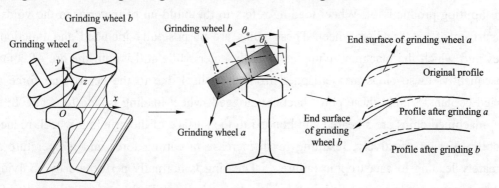

Figure 3.7 Rail grinding mechanism

The grinding mode of a grinding train is composed of different values of travel speed v of a grinding train, grinding angle α and grinding press p. During the grinding construction process, a differential grinding mode is selected to control the final grinding profile obtained based on the actual damage and wear status of the on-site rail.

3.4.4 Best Practice in Aerospace Grinding

The requirements for high productivity and accuracy for the machining of components in the aerospace industry has led to significant advancements in grinding technology. A number of different grinding techniques developed for, but not exclusively used, in this sector were evaluated to assess their applicability for rail grinding operations [2]. These are detailed as follows.

(1) Creep feed grinding

Creep feed grinding (CFG) was a concept introduced in the 1960s that extended grinding from a surface finishing operation into a stock removal process. The process is very different from conventional surface grinding and is characterized by a large depth of cut applied at a slower work piece feed rate. The benefits of the process include a low force per grinding grit, which provides a low force for a given area of cut, meaning large depths of material can be removed in one cutting pass. However, friction is high, producing high contact zone temperatures and high energy requirements in the process. The large contact zone allows slow thermal dissipation; therefore, reducing potential damage to the surface, however, typically a large amount of cutting fluid is applied to transfer the heat away from the cut zone.

(2) Continuous dressing

A dressing process in grinding is a method of preparing the cutting profile and adjusting the cutting grain sharpness to provide optimum cutting conditions. A process of continuous dressing (CD) was applied to CFG when problems with wheel clogging and loss of profile were limiting productivity. Wheel loading refers to the build-up of material in the voids on the surface of the grinding wheel. These voids are purposefully included and function as pores into which the machined chip can initially enter while still within the arc of contact. Subsequently, these chips are jettisoned from the wheel due to the centrifugal force and applied coolant. Should these pores become clogged, wheel loading is said to occur. In this case, machined chips are smeared and bonded to the surface of the grinding wheel, reducing its ability to cut effectively; resulting in an increase in cutting force and temperature and ultimately leading to catastrophic failure [3]. Dressing is normally performed by applying a harder abrasive material, e.g. diamond, to a rotating grinding wheel to remove the top surface and expose fresh cutting grains. The benefits for this type of process include: increased consistency of cutting conditions, controlled wheel wear and maintenance of the profile geometry of the wheel. In addition, it has been shown that the application of CD during grinding can reduce the specific grinding energy by a factor of four during cutting.

(3) VIPER grinding

VIPER grinding is a concept developed and patented by Rolls-Royce PLC. It uses CFG with an improved coolant application technique. Coolant is applied as a jet with high flow rate and pressure at a position on the grinding wheel just before the cut zone. This has the effect of cleaning the wheel and thus preventing wheel clogging, and allowing coolant to penetrate into the grinding wheel, thereby providing increased lubrication and cooling.

VIPER grinding offers an increased performance in productivity through its cleaning and cooling capability allowing a larger MRR (maintenance repairs replacements) without damaging the work piece material. To implement this requires the application of coolant just before the cut zone at a 70 bar pressure and a flow rate of approximately 100 l/min. It is understood that, for a rail grinding setup, all coolant products will end up on the track and thus they have to be in conformance with environmental standards. It is possible to meet this requirement, but of course, the coolant still has to be carried on the train.

(4) High speed grinding

High speed grinding (HSG) is defined by Tawakoli as grinding wheel surface speeds exceeding 60 ~ 80 m/s. This is the common definition for HSG and is different from the patented train technology described in the literature. The exact value of the required wheel surface speed to achieve HSG is not defined; however, the majority of traditional grinding processes operate up to 50 m/s. As a result, HSG is classed as any operation where

Figure 3.8　A steel core plated grinding wheel

the wheel surface speed exceeds this value. The main advantage of HSG is that higher wheel speeds reduce the size of the metal chip produced by each abrasive grain during cutting. This allows for higher productivity due to increased work piece feed rates while maintaining a chip size that the grinding wheel can accommodate. In general, increasing the wheel speed is beneficial to any grinding process. The main technical issues occur with the tooling. At high wheel speeds, the wheels must be rotated at high speeds. This can result in wheels bursting. Steel core wheels, as shown in Figure 3.8，are required to avoid this phenomenon.

High speed rail grinding

(5) High efficiency deep grinding

High efficiency deep grinding (HEDG) is a novel abrasive machining process that readily achieves very high MRR values. The technique is a development of CFG and HSG. By combining a high depth of cut, work piece and wheel surface speeds, the cutting conditions, and hence the thermal, behavior, of the process are fundamentally changed. These changes in thermal behavior result in high temperatures in the wheel-work piece contact zone (contact layer), which facilitates material removal. However, due to the high work piece

Rail grinder under train view

speed the heat is removed with the grinding chip before significant penetration into the work piece surface is allowed to occur. It was decided that this would be the most suitable technology to explore for a potential increase in rail grinding performance, and from this point onwards all modelling work was focused on this technology. This selection was made as HEDG technology has the greatest potential to deliver increased grinding train speeds via an increase in the MRR, and the primary focus of this work is the improvement of preventative re-profiling operations. Additionally, it was also felt that it is imperative to avoid a heat-affected zone on the rail head, and HEDG is the only process that combines high productivity with minimization of work piece heating [4].

Words and Expressions

abrasive [əˈbreɪsɪv] a substance that abrades or wears down	*adj.* 有磨蚀作用的；摩擦的；粗糙的
aerospace industry [ˈeərəʊspeɪs ˈɪndəstri]	航空航天工业
centrifugal [ˌsentrɪˈfjuːgl] tending to move away from a center	*adj.* 离心的
clog [ˈklɒgɪŋ] to block sth.	*v.* 堵塞
coolant [ˈkuːlənt] a fluid agent (gas or liquid) that produces cooling	*n.* 冷冻剂，冷却液，散热剂
cut depth [kʌt depθ]	切深
exceed [ɪkˈsiːd] go beyond	*v.* 超过；超越
heat-affected zone [hiːt əˈfektɪd zəʊn]	热影响区
infrastructure [ˈɪnfrəstrʌktʃə(r)] the basic structure or features of a system or organization	*n.* 基础设施；基础建设
metal chip [ˈmetl tʃɪp]	切屑
non-driven [nɒn ˈdrɪvən]	非驱动的
peripheral grinding [pəˈrɪfərəl ˈgraɪndɪŋ]	外圆磨削
potential damage [pəˈtenʃəl ˈdæmɪdʒ]	潜在损害
preventative [prɪˈventətɪv] intended to stop sth. you do not want to happen	*adj.* 预防性的
rail grinding [reɪl ˈgraɪndɪŋ]	钢轨打磨

sector ['sektə(r)] a part of an area of activity	n. 领域
steel core wheel [stiːl 'kɔː wiːl]	钢芯轮
target profile ['tɑːɡɪt 'prəʊfaɪl]	目标特征；目标轮廓
thermodynamics [ˌθɜːməʊdaɪ'næmɪks] the branch of physics concerned with the conversion of different forms of energy	n. 热力学
wear status [weə 'steɪtəs]	磨损状态

Notes

[1] 本句可译为：英国的基础设施更复杂，具有大量的弯曲部分，并且需要根据轨道使用情况变化切深要求。

[2] 本句可译为：评估了许多为该行业开发但不专门使用的不同的磨削技术，从而评价这些磨削技术在钢轨磨削作业中的适用性。

[3] 本句可译为：在这种情况下，加工产生的切屑被粘擦到砂轮的表面，降低了其有效切割的能力；导致切削力和温度的增加，并最终导致严重的失效。

[4] 本句可译为：此外，也有必要避免在轨头上产生热影响区，而 HEDG 是将高生产率与工件产生的加工热量最小化结合的唯一工艺方法。

Questions for Discussion

1. What problems can rail grinding solve?
2. What are the key problems in rail grinding technology?
3. In Section 3.4.4, several grinding techniques are introduced, please summarize their technical characteristics respectively.

3.5　Micromachine and Nanomachine

3.5.1　Micromachine

From the beginning, mankind seems instinctively to have desired large machines and small machines. That is, "large" and "small" in comparison with human-scale. Machines larger than human are powerful allies in the battle against the fury of nature; smaller machines are loyal partners that do whatever they are told [1].

If we compare the facility and technology of manufacturing larger machines, common sense tells us that the smaller machines are easier to make. Nevertheless, throughout the history of technology, larger machines have always stood out. The size of the restored models of the water-mill invented by Vitruvius in the Roman Era, the windmill of the Middle Ages, and the steam engine invented by Watt is overwhelming. On the other hand, smaller machines in history of technology are mostly tools. If smaller machines are easier to make, a variety of such machines should exist, but until modern times, no significant small machines existed except for guns and clocks.

This fact may imply that smaller machines were actually more difficult to make. Of course, this does not mean simply that it was difficult to make a small machine: It means that it was difficult to invent a small machine that would be significant to human being.

Some people might say that mankind may not have wanted smaller machines. This theory, however, does not explain the recent popularity of palm-size mechatronics products.

The absence of small machines in history may be due to the extreme difficulty in manufacturing small precision part.

(1) History of micromachine

The dream of the ultimate small machine, or micromachine, was first depicted in detail in the 1966 movie "Fantastic Voyage". At the time the study of micromachining of semiconductors had already begun. Therefore, manufacturing minute mechanisms through micromachining of semiconductors would have been possible, even at that time. There was, however, a wait of over 20 years before the introduction of electrostatic motors and gears made by semiconductor micromachining.

Why didn't the study of micromachining and the dream of micromachines meet earlier? A possible reason for this is as follows. In addition to micromachining, the development of micromachines requires a number of technologies including materials, instrumentation, control, energy, information processing, and design. Before micromachine research and development can be started all of these technologies must reach a certain level. In other words, the overall technological level, as a whole, must reach a certain critical point, but it hadn't reached that point then.

Approximately 20 years after "Fantastic Voyage", the technology level for micromachines finally reached a critical point. Micromotors and microgears made by semiconductor micromachining were introduced at about that time triggering the development of micromachines.

(2) Micromachine as gentle machines

How do micromachines of the future differ from conventional machines? How will they change the relationship between nature and humans?

The most unique feature of a micromachine is, of course, its small size. Utilizing its tiny dimensions, a micromachine can perform tasks in a revolutionary way that would be impossible for conventional machines. That is, micromachines do not affect the object or the environment as much as conventional machines do Micromachines perform their tasks gently. This is a fundamental difference between micromachines and conventional machines.

The medical field holds the highest expectations for benefits from this feature of micromachines. Diagnosis and treatment will change drastically from conventional methods, and "Fantastic Voyage" may no longer be a fantasy. If a micromachine can gently enter a human body to treat illnesses humans will be freed from painful surgery and uncomfortable gastro-camera testing. Furthermore, if micromachines can halt the trend of ever-increasing size in medical equipment, it could slow the excess growth and complexity of medical technology, contributing to the solving of serious problems with high medical costs for citizens.

Micromachines are gentle also in terms of machine maintenance, since they can be inspected and repaired without difficulty. The more complex the machine the more susceptible it is to malfunction due to overhaul and assembly. In addition, there have been more instances of human errors during overhaul and assembly. It is good for the machine if overhaul is not necessary. It is even better if maintenance can be performed without stopping the machine. Repeated stop-and-go operation will accelerate damage of the machine due to excess stress caused by thermal expansion.

Such gentleness of a micromachine is an advantage, as well as a weakness in that a micromachine is too fragile to resist the object or the environment. This is the drawback of the microscale objects.

For example, a fish can swim freely against the current, but a small plankton cannot. This is result of physical laws and nothing can change it. Still the plankton can live and grow in the natural environment by conforming to the environment.

Unlike conventional machines which fight and control nature, micromachines will probably adapt to and utilize nature. If a micromachine cannot proceed against the current, a way will be found to proceed with the flow, naturally avoiding collisions with obstacles.

(3) Microelectronics and mechatronics

The concept of micromachines and related technologies is still not adequately unified, as these are still at the development stage. The micromachines and related technologies are currently referred to by a variety of different terms. In the United States, the accepted term is "Micro Electro Mechanical Systems" (MEMS). In Europe, the term Microsystems Technology (MST) is common, while the term "microengineering" is sometimes used in Britain. Meanwhile in Australia "micromachine". The most common term if it is translated into English is "micromachine" in Japan. However, "microrobot" and "micromechanism" are also available case by case.

The appearance of these various terms should be taken as reflecting not merely diversity of expression, but diversity of the items referred to pending on whether the item referred to is an object or a technology, the terminology may be summed up as follows:
- Object: microrobot, micromechanism;
- Technology: microengineering, MST;
- Object technology: MEMS, micromachine.

With regard to technology, if we summarize the terms according to where the technology for micromachine systems branched off from, and whether the object dealt with by the technology in question is an element or a machine system, the terms can be organized as follows. That is, MEMS and MST stem from mechatronics, and have developed dealing mainly with machine systems. In this sense, MEMs and MST on the one hand and micromachines and microengineering on the other hand form two separate groups, but as the former has started to move in the direction of machine systems, while the latter has already incorporated microelectronics, the differences between the two groups are gradually disappearing.

Looking at the areas included in the two groups, given that the machine systems which are the main concern of micromachines include elements, and given also that micromachines include microelectronics, it would be natural to assume that micromachines include MEMS and MST.

(4) The evolution of machines and micromachines

Many researchers see micromachines as the ultimate in mechatronics developed out of machine systems.

Ever since the Industrial Revolution, machine systems have grown larger and larger in the course of their evolution. Only very recently has evolution in the opposite direction begun, with the appearance of mechatronics. Devices such as video cameras, tape recorders, portable telephones, portable copiers which at one time were too large to put one's arms around, now fit on the palm of one's hand.

Miniaturization through mechatronics has resulted mainly from the development of electronic controls and control software for machine systems, but the changes to the structural parts of machine systems have been minor compared to those in the control systems.

The next target in miniaturization of machine systems is miniaturization of the structural parts left untouched by present mechatronics. These are the micromachines which are seen as the ultimate in mechatronics.

Seen in this light the aim of micromachines can be expressed as "Micromachines are autonomous machines which can be put on a fingertip, composed of parts the smallest sized of which is a few dozen micrometers."

That is, since micromachines which can be put on a fingertip have to perform operations in spaces inaccessible to humans, they are required to be autonomous and capable of assessing situations independently, as are intelligent robots. To achieve this kind of functionality, a large number of parts must be assembled in a confined space. This factor determines the size of the smallest parts, and given the resolution of micromachining systems. A target size of several dozen micrometers should be achievable.

Micromachines are unconventional artifacts with respects to their gentle features to people and nature. The current diversity of the definition of them are originated from development objectives and technological starting points. Micromachine technologies, in view of their development prospect, are expected as generic technologies for the twenty-first century to support industry and medicine as well as daily life. Micromachine technologies are essential also for improving the conventional machines in general.

3.5.2 Nanomanufacturing

(1) Nanotechnology

Nanotechnology is the understanding and control of matter at the nanoscale, at dimensions between approximately 1 and 100 nm, where unique phenomena enable novel applications. Encompassing nanoscale science, engineering, and technology, nanotechnology involves imaging, measuring, modeling, and manipulating matter at this length scale.

Matter such as gases, liquids, and solids can exhibit unusual physical, chemical, and biological properties at the nanoscale, differing in important ways from the properties of bulk materials and single atoms or molecules [2]. Some nanostructured materials are stronger or have different magnetic properties compared to other forms or sizes or the same material. Others are better at conducting heat or electricity. They may become more chemically reactive or reflect light better or change color as their size or structure is altered.

(2) Manufacturing at the nanoscale

Manufacturing at the nanoscale is known as nanomanufacturing. Nanomanufacturing involves scaled-up, reliable, and cost-effective manufacturing of nanoscale materials structures, devices, and systems. It also includes research, development, and integration of top-down processes and increasingly complex bottom-up or self-assembly processes.

In more simple terms, nanomanufacturing leads to the production of improved materials and new products. There are two basic approaches to nanomanufacturing, either top-down or bottom-up.

1) Top-down approaches

Top-down fabrication reduces large pieces of materials all the way down to the nanoscale like someone carving a model airplane out of a block of wood. This approach requires larger amounts of materials and can lead to waste if excess material is discarded. The removing methods can be mechanical, chemical, and electrochemical, etc., depending on the material of the base substrate and requirement of the feature sizes. The formed structures usually share the same material with the base substrate.

There are a couple of manufacturing technologies in the conventional scale which can be categorized "top-down". Milling is a representative example. In the milling process material is selectively removed from the substrate, usually a metal sheet, forming a cavity with certain geometries. The dimensions of the cavity depend on the travel path mill, which can be precisely controlled with the help of computer-assisted numerical systems.

People have attempted to extend "top-down" method into the nanometer domain, and supplemented the mechanical removing methods with chemical and electrochemical methods. Patterning (using photolithography or electron beam lithography) is used to etch away the material, as in building integrated circuits.

2) Bottom-up approaches

As the opposite to "top-down" fabrication technologies, "bottom-up" methods refer to a set of technologies which fabricate by stacking materials on top of a base substrate. These methods are similar in principle to welding and riveting at the conventional scale, in which a different type of material is attached to the base component by melted solder or physical fitting. In welding and riveting, attention is mainly paid to the strength of the contact area in order to maintain the construct as a reliable component for high load application.

Similarly, in "bottom-up" nanofabrication, people can create products by building them up from atomic- and molecular-scale components, which can be time-consuming. The adhesion of the surface layer to the base substrate is also an important concern. There is extensive research on the surfactants to enhance adherence and avoid cracks during the subsequent processing. Research has also focused on autonomous patterning of the surface layer into nanometer scale features since manipulation of nanoscale components is not ever an easy task, as compared to that at the conventional scale.

Scientists are exploring the concept of placing certain molecular-scale components together that will spontaneously "self-assemble", from the bottom up into ordered structures.

Self-assembly is the assembly of molecules without guidance or management from an outside source. Generally speaking, it refers to intermolecular self-assembly, which aggregates individual molecules with desired patterns to form molecular assemblies. The assemblies are normally built on top of a solid surface and patterned in desired structures with nanometer precision. Nanofabrication methods based on self-assembly provide a new strategy for the successful fabrication of next generation devices. They offer the promise of nanostructured fabrication of devices for novel electronic applications such as quantum computing, sensing, and integration with biotechnology.

(3) Nanomanufacturing

Within the top-down and bottom-up categories of nanomanufacturing, there are a growing number of new processes that enable nanomanufacturing. Among these are:

- Chemical vapor deposition is a process in which chemicals react to produce very pure, high-performance films.
- Molecular beam epitaxy is one method for depositing highly controlled thin films.
- Atomic layer epitaxy is a process for depositing one-atom-thick layers on a surface.
- Dip pen lithography is a process in which the tip of an atomic force microscope is "dipped" into a chemical fluid and then used to "write" on a surface, like an old fashioned ink pen onto paper.
- Nanoimprint lithography is a process for creating nanoscale features by "stamping" or "printing" them onto a surface.
- Roll-to-roll processing is a high-volume process to produce nanoscale devices on a roll of ultrathin plastic or metal.
- Self-assembly describes the process in which a group of components come together to form an ordered structure without outside direction.

Structures and properties of materials can be improved through these nanomanufacturing processes. Such nanomaterials can be stronger, lighter, more durable water-repellent, anti-reflective, self-cleaning, ultraviolet- or infrared-resistant, antifog, antimicrobial, scratch-resistant, or electrically conductive, among other traits.

(4) Applications of nanomanufacturing

There is an all-pervading trend to higher precision and miniaturization, and to illustrate this a few applications will be briefly referred to in the fields of mechanical engineering, optics and electronics. It should be noted however that the distinction between mechanical engineering and optics is becoming blurred, now that machine tools such as precision grinding machines and diamond-turning lathes are being used to produce optical components, often by personnel with a background in mechanical engineering rather than optics [3]. By a similar token, mechanical engineering is also beginning to encroach on electronics particularly in the preparation of semiconductor substrates.

1) Mechanical engineering

One of the earliest applications of diamond turning was the machining of aluminum substrates for computer memory discs, and accuracies are continuously being enhanced in order to improve storage capacity: surface finishes of 3 nm are now being achieved. In the related technologies of optical data storage and retrieval, the tolerances of the critical dimensions of the disc and reading head are about 0.25 μm. The tolerances of the component parts of the machine tools used in their manufacture, i. e. the slide ways and bearings fall well within the nanotechnology range.

Some precision components falling in the manufacturing tolerance band of 5~50 nm include gauge blocks, diamond indenter tips, microtome blades Winchester disc reading heads and ultra precision XY tables (Taniguchi 1986). Examples of precision cylindrical components in two very different fields, and which are made to tolerances of about 100 nm, are bearings for mechanical gyroscopes and spindles for video cassette recorders.

The theoretical concept that brittle materials may be machined in a ductile mode has been known for some time. If this concept can be applied in practice it would be of significant practical importance because it would enable materials such as ceramics, glasses and silicon to be machined with minimal sub-surface damage, and could eliminate or substantially reduce the need for lapping and polishing.

Typically, the conditions for ductile-mode machining require that the depth of cut is about 100 nm and that the normal force should fall in the range of 0.1~0.01 N. These machining conditions can be realized only with extremely precise and stiff machine tools, and with which quartz has been ground to a surface roughness of 2 nm peak-to-valley. The significance of this experimental result is that it points the way to the direct grinding of optical components to an optical finish. The principle can be extended to other materials of significant commercial importance, such as ceramic turbine blades which at present must be subjected to tedious surface finishing procedures to remove the structure-weakening cracks produced by the conventional grinding process.

2) Optics

In some areas in optics manufacture there is a clear distinction between the technological approach and the traditional craftsman's approach, particularly where precision machine tools are employed. On the other hand, in lapping and polishing, there is a large grey area where the two approaches overlap. The large demand for infrared optics from the 1970s onwards could not be met by the traditional suppliers, and provided a stimulus for the development and application of diamond-turning machines to optic manufacture. The technology has now progressed and the surface figure and finishes that can be obtained span a substantial proportion of the nanotechnology range important applications of diamond-turned optics are in the manufacture of unconventionally shaped optics, for example axicons and more generally, aspherics and particularly offaxis components, such as paraboloids.

The mass production (several million per annum) of the miniature aspheric lenses used in compact disc players and the associated lens moulds provides a good example of the merging of optics and precision engineering. The form accuracy must be better than 0.2 μm and the surface roughness must be below 20 nm to meet the criterion for diffraction limited performance.

3) Electronics

In semiconductors, nanotechnology has long been a feature in the development of layers parallel to the substrate and in the substrate surface itself, and the need for precision is steadily increasing with the advent of layered semiconductor structures [4]. About one quarter of the entire semiconductor physics community is now engaged in studying aspects of these structures. Normal to the layer surface, the structure is produced by lithography, and for research purposes at least, nanometer-sized features are now being developed using X-ray and electron and ion-beam techniques.

Words and Expressions

assembly [əˈsembli] a group of machine parts that fit together to form a self-contained unit	n. 装配
atomic layer epitaxy [əˈtɒmɪk ˈleɪə ˈepɪtæksi]	原子层外延
biotechnology [ˌbaɪəʊtekˈnɒlədʒi] the branch of molecular biology that studies the use of microorganisms to perform specific industrial processes	n. 生物工艺学；生物技术
bulk materials [bʌlk məˈtɪərɪəlz]	散装物料；块材
cost-effective [kɒst ɪˈfektɪv]	adj. 有成本效益的；划算的
device [dɪˈvaɪs] an object or a piece of equipment that has been designed to do a particular job	n. 装置；设备
dip pen lithography [dɪp pen lɪˈθɒɡrəfi]	蘸笔光刻
electrostatic [ɪˌlektrəʊˈstætɪk] concerned with or producing or caused by static electricity	adj. 静电的
fury [ˈfjʊəri] a feeling of intense anger	n. 狂怒；暴怒
gastro-camera [ɡæstrəʊ ˈkæmərə]	n. 胃内照相机

gear [gɪə(r)] a toothed wheel that engages another toothed mechanism in order to change the speed or direction of transmitted motion	n. 齿轮；排挡
generic technology [dʒəˈnerɪk tekˈnɒlədʒi]	共通技术
gyroscope [ˈdʒaɪrəskəʊp] rotating mechanism in the form of a universally mounted spinning wheel that offers resistance to turns in any direction	n. 陀螺仪；回转仪
human-scale [ˈhjuːmən ˈskeɪl]	人体尺度
instrumentation [ˌɪnstrəmenˈteɪʃən] an artifact (or system of artifacts) that is instrumental in accomplishing some end	n. 使用仪器；装设仪器
ion-beam [ˈaɪən biːm]	离子束
machine maintenance [məˈʃiːn ˈmeɪntənəns]	机器保养
malfunction [ˌmælˈfʌŋkʃən] fail to function normally	v. 失灵；发生故障
mechatronic [mekæˈtrɒnɪk] integration of mechanical and electric	n. 机电一体化
microscale [ˈmaɪkrəʊskeɪl]	adj. 微尺度的
miniaturization [ˌmɪnətʃəraɪˈzeɪʃən] act of making on a greatly reduced scale	n. 小型化；缩形技术
molecular beam epitaxy (MBE) [məˈlekjələ biːm ˈepɪtæksi]	分子束外延
nanofabrication [ˌnænəʊfæbrɪˈkeɪʃən] the branch of fabrication	n. 纳米制造；纳米加工
nanoimprint lithography [ˌnænəʊɪmˈprɪnt lɪˈθɒgrəfi]	纳米压印光刻
nanoscale [ˈnænəʊskeɪl] a peer-reviewed scientific journal covering experimental and theoretical research in all areas of nanotechnology and nanoscience	n. 纳米级
overhaul [ˈəʊvəhɔːl] periodic maintenance on a car or machine	n. 大修；检查；彻底检修；详细检查
photolithography [ˌfəʊtəlɪˈθɒgrəfi] a planographic printing process using plates made from a photographic image	n. 影印石版术；照相平版

plankton [ˈplæŋktən] the aggregate of small plant and animal organisms that float or drift in great numbers in fresh or salt water	*n.* 浮游生物
quantum computing [ˈkwɒntəm kəmˈpjuːtɪŋ]	量子计算
roll-to-roll processing [rəʊl tuː rəʊl ˈprʊsesɪŋ]	滚轧加工
self-assembly process [self əˈsembli ˈprəʊses]	自发装配过程
semiconductor [ˌsemɪkənˈdʌktə(r)] a conductor made with semiconducting material	*n.* 半导体
structure-weakening crack [ˈstrʌktʃə ˈwiːkənɪŋ kræk]	结构弱化裂纹
trigger [ˈtrɪɡə(r)] to make sth. happen suddenly	*n.* 起动；触发；控制

Notes

[1] 本段可译为：从一开始，人类似乎本能地就有一种想制造"大机器"和"小机器"的愿望。这里的所谓"大"和"小"是对于人类身体本身的尺寸而言的。比人体大的机器将成为人类同暴虐无情的自然界作斗争的得力帮手，而那些小机器则只能乖乖听从人类的命令，让干什么就干什么。

[2] 本段可译为：物质，如气体、液体和固体，在纳米尺度上可以表现出不寻常的物理、化学和生物特性，与块状材料和单个原子或分子的特性有很重要的区别。

[3] 本段可译为：然而，应当指出的是，机械工程和光学的区别正在变得模糊，现在诸如精密磨床和金刚石车床之类的机床通常由具有机械工程背景而非光学专业的人员用于生产光学元件。

[4] 本段可译为：在半导体技术中，纳米技术同半导体基片材料和平行于基片的各层材料的生长密切相关，而且随着半导体层状结构的出现，对精度的要求愈来愈高。

Questions for Discussion

1. Describe the main approaches that are used in nanotechnology.
2. Choose an application of nanotechnology from a newspaper, magazine, professional journal or website, clearly and briefly summarize it.
3. What are the main links between nanotechnology and micromachine?

Chapter 4
Mechanical Design and Theory

4.1 Problems in Mechanical Design

Different types of mechanical design problems

For a new product, there will be a lot of original design work to be done. As the design process proceeds, we will configure the various parts. To determine the thickness of the frying surface we will analyze the heat conduction of the frying component, which is parametric design. And we will select a heating element and various fasteners to hold the components together. Further, if we are clever, we may be able to redesign an existing product to meet some or all of the requirements. Each of the italicized terms is a different type of design problem. It is rare to find a problem that is purely one type.

4.1.1 Selection Design

Selection design involves choosing one item (or maybe more) from a list of similar items. We do this type of design every time we choose an item from a catalog. It may sound simple, but if the catalog contains more than a few items and there are many different features to the items, the decision can be quite complex.

To solve a selection problem we must start with a clear need. The catalog or the list of choices then effectively generates potential solutions for the problem. We must evaluate the potential solutions with respect to our specific requirements to make the right choice. Consider the following example. During the process of designing a product, an engineer must select a bearing to support a shaft. The known information is shown in Figure 4.1. The shaft has a diameter of 20 mm (0.787 in). There is a radial force of 6675 N on the shaft at the bearing, and the shaft rotates at a maximum of 2000 rpm. The housing to support the bearing is still to be designed. All we need to do is select a bearing to meet the needs. The information on shaft size, maximum radial force, and maximum rpm given in bearing catalogs enables us to quickly develop a list of potential bearings (shown in Table 4.1). This is the simplest type of design problem we could have, but it is still incompletely defined. We do not have enough information to select among the five possible choices. Even if a short list is developed—the most likely candidates being the 42-mm-deep groove ball bearing and the 24-mm needle bearing—there is no way to make a good decision without more knowledge of the function of the bearing and of the engineering requirements on it.

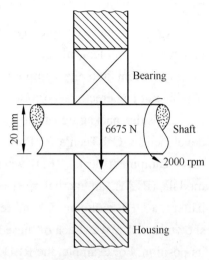

Figure 4.1　Load on a shaft

Table 4.1　Potential bearings for a shaft

Type		Outside diameter(mm)	Width(mm)	Load rating(lb)	Speed limit(rpm)	Catalog number
Deep-groove ball bearing		42	8	1,560	18,000	6,000
		47	14	2,900	15,000	6,204
		52	15	3,900	9000	6,304
Angular-contact ball bearing		47	14	3,000	13,000	7,204
		37	9	1,960	34,000	71,904
Roller bearing		47	14	6,200	13,000	204
		52	15	7,350	13,000	220
Needle bearing		24	20	1,930	13,000	206
		26	12	2,800	13,000	208
Nylon bushing		23	Variable	290 ⋮ 8	10 ⋮ 500	4,930

4.1.2 Configuration Design

A slightly more complex type of design is called configuration or packaging design. In this type of problem, all the components have been designed and the problem is how to assemble them into the completed product. Essentially, this type of design is similar to playing with an erector set or other construction toy, or arranging living-room furniture.

Consider packaging of the assemblies in the MER, the NASA mars exploration rover developed by Cal Tech's Jet Propulsion Laboratory. The body of the MER is made up of a rover equipment deck (RED) where all the experiments are mounted, a rover electronics module (REM), an inertial measurement unit (IMU), a warm electronics box (WEB), a battery, a UHF radio, an X-band telecom HW, and a solid-state power amplifier (SSPA), as shown in Figure 4.2. Each of these assemblies is of known size and has certain constraints on its position. For example, the RED must be on top and the WEB on the bottom, but many of the other major assemblies can be anywhere inside the envelop defined by these two.

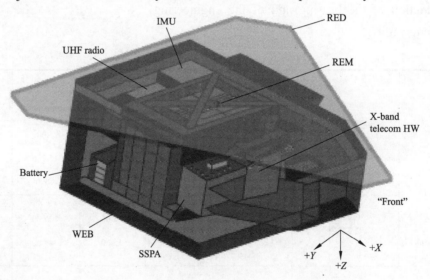

Figure 4.2   The major assemblies in the MER

Configuration design answers the question, how do we fit all the assemblies in an envelop or where do we put what. One methodology for solving this type of problem is to randomly select one component from the list and position it so that all the constraints on that assembly are meet. We could start with the REM in the middle, then we select and place a second component. This procedure is continued until either we run into a conflict or all the components are in the MER. If a conflict arises, we back up and try again. For many configuration problems, some of the components to be fitted into the assembly can be altered in size, shape, or function, giving the designer more latitude to determine potential configurations and making the problem solution more difficult.

4.1.3 Parametric Design

Parametric design involves finding values for the features that characterize the object being studied. This may seem easy enough—just find some values that meet the requirements. However, consider a very simple example. We want to design a cylindrical storage tank that must hold 4 m³ of liquid. This tank is described by the parameters r, its radius, and l, its length, and its volume is determined by

$$V = \pi r^2 l \tag{4.1}$$

Given a volume equal to 4 m³, then

$$r^2 l = 1.273 \tag{4.2}$$

We can see that an infinite number of values for the radius and length will satisfy this equation. To what values should the parameters be set? The answer is not obvious, nor even completely defined with the information given.

Let us extend the concept further. It may be that instead of a simple equation, a whole set of equations and rules govern the design. Consider the instance in which a major manufacturer of copying machines had to design paper-feed mechanisms for each new copier (A paper feed is a set of rollers, drive wheels, and baffles that move a piece of paper from one location to another in the machine). Many parameters—the number of rollers, their positions, the shape of the baffles, and the like—characterize this particular design problem, but obviously there are certain similarities in paper feeders, regardless of the relative positions of the beginning and end points of the paper, the obstructions (other components in the machine) that must be cleared, and the size and weight of the paper. The company developed a set of equations and rules to aid designers in developing workable paper paths, and using this information, the designers could generate values for parameters in new products.

4.1.4 Original Design

Any time the design problem requires the development of a process, assembly, or component not previously in existence it calls for an original design (It can be said that if we have never seen a wheel and we design one, then we have an original design). Though most selection, configuration, and parametric problems are represented by equations, rules, or some other logical scheme, original design problems usually cannot be reduced to any algorithm. Each one represents something new and unique.

In many ways the other types of design problems—selection, configuration, and parametric—are simply constrained subsets of an original design. The potential solutions are limited to a list, an arrangement of components, or a set of related characterizing values. Thus, if we have a clear methodology for performing original design, we should be able to solve any design problem with a more limited set of potential solutions.

4.1.5 Redesign

Most design problems solved in industry are for the redesign of an existing product. Suppose a manufacturer of hydraulic cylinders makes a product that is 0.25 m long. If the customer needs a cylinder 0.3 m long, the manufacturer might lengthen the outer cylinder and the piston rod to meet this special need. These changes may require only parameter changes, or they may require something more extensive. What if the materials are not available in the needed length, or cylinder fill time becomes too slow with the added length? Then the redesign effort may require much more than parameter changes. Regardless of the change, this is an example of redesign, the modification of an existing product to meet new requirements.

Many redesign problems are routine; the design domain is so well understood that the method used can be put in a handbook as a series of formulas or rules. The parameter changes in the example of the hydraulic cylinder are probably routine for the manufacturer.

The hydraulic cylinder can also be used as an example of a mature design, in that it has remained virtually unchanged over many years. There are many examples of mature designs in our everyday lives: pencil sharpeners, hole punches, and staplers are a few found on the average desk. For these products, the knowledge about the design problem is high. There is little more to learn.

However, consider the bicycle. The basic configuration of the bicycle—the two tensioned, spoked wheels of equal diameter, the diamond-shaped frame, and the chain drive—was fairly refined late in the nineteenth century. While the 1890 Humber bicycle, shown in Figure 4.3, looks much like a modern bicycle, not all bicycles of this era were of this configuration. The Otto dicycle, shown in Figure 4.4, had two spoked wheels and a chain; stopping and steering this machine must have been a challenge. In fact, the technology of bicycle design was so well developed by the end of the nineteenth century that a major book on the subject, *Bicycles and Tricycles: An Elementary Treatise on Their Design and Construction*, was published in 1896. The only major change in bicycle design since the publication of that book was the introduction of the derailleur in the 1930s.

Figure 4.3 1890 Humber bicycle

Figure 4.4 Otto dicycle

However, in the 1980s the traditional bicycle design began to change again. For example, the mountain bike shown in Figure 4.5 no longer has a diamond-shaped frame. Why did a mature design like a bicycle begin evolving again? First, customers are always looking for improved performance. Bicycles of the style shown in Figure 4.5 are better able to handle rough terrain than traditional bikes. Second, there is improved understanding of human comfort, ergonomics, and suspensions. Third, customers are always looking for something new and exciting even if performance is not greatly improved. Fourth, materials and components have improved.

Figure 4.5　Mountain bike

The point is that even mature designs change to meet new needs, to attract new customers, or to take advantage of new materials. Part of the design of a new bicycle like the Marin Mount Vision is routine, and part is original. Additionally, many sub-problems were parametric problems, selection problems, and configuration problems. Thus, the redesign of a product, even a mature one, may require a wide range of design activity[1].

4.1.6　Variant Design

Sometimes companies will produce a large number of variants as their products. A variant is a customized product designed to meet the needs of the customer. For example, when you order a new computer from companies such as Dell, you can specify one of three graphics cards, two battery configurations, three communication options, and two levels of memory. Any combination of these is a variant that is specifically tuned to your needs. Also, Volvo trucks estimates that of the 50,000 parts it has in its inventory it annually supplies over 5,000 variants, different truck models specifically assembled to meet the needs of the customer.

4.1.7 Conceptual Design and Product Design

These are catchall terms for two parts of the product development process. First, you must develop a concept and then refine the concept into a product. The activities during the conceptual and product development phases may make use of original, parametric, and selection design and redesign as needed.

Words and Expressions

Parametric design [ˌpærəˈmetrɪk dɪˈzaɪn]	参数设计
bicycle [ˈbaɪsɪkl] a wheeled vehicle that has two wheels and is moved by foot pedals	n. 二轮（车）

Notes

[1] 本句中间有插入短语。

Questions for Discussion

Decompose a simple system such as a home appliance, bicycle, or toy into its assemblies, components, electrical circuits, and the like.

4.2 Machine Elements

Machine elements in mechanical design

Mechanical design and theory deals with the concepts, procedures, data, and decision analysis techniques necessary to design machine elements commonly found in mechanical devices and system. One completing a course of study of the fundamentals of mechanical design and theory should be able to execute original designs for machine elements and integrate the elements into a system composed of several elements.

This process requires a consideration of the performance requirements of an individual element and of the interfaces between elements as they work together to form an integrated system. For example, a gear must be designed to transmit power at a given speed. The design must specify the number of teeth, pitch, tooth form, face width, pitch diameter, material, and method of heat treatment, etc. But the gear design also affects, and is influenced by the mating gear, the shaft carrying the gear, and the environment in which it is to operate. Furthermore, the shaft must be supported by bearings, which must be contained in a housing. Thus, the designer should keep the complete system in mind while designing each individual element. The machine elements commonly found in mechanical devices include the following parts: the belt drives and chain drives, the gears, the wormgear drives, keys, couplings, the shaft, the bearings (the rolling bearing / the plain surface bearing), the screws, the fasteners, the shafts, etc.

4.2.1 Belt Drives and Chain Drives

Belts and chains represent the major types of flexible power transmission elements. Usually, rotary power is developed by the electric motor, but motors typically operate at too high a speed and deliver too low a torque to be appropriate for the final drive application. For a given power transmission, the torque is increased in proportion to the amount that rotational speed is reduced. So some speed reduction is often desirable. The high speed of the motor makes belt drives somewhat ideal for that first stage of reduction. A smaller drive pulley is attached to the motor shaft, while a larger diameter pulley is attached to a parallel shaft that operates at a correspondingly lower speed. At the low speed, high torque condition, chain drives become desirable. The high torque causes high tensile forces to be developed in the chain. The elements of the chain are typically metal, and they are sized to withstand the high forces. The links of chains are engaged in toothed wheels called sprockets to provide positive mechanical drive, desirable at the low speed, high torque conditions.

A belt is a flexible power transmission element that seats tightly on a set of pulleys or sheaves. Figure 4.6 shows the basic layout. When the belt is used for speed reduction, the typical case, the smaller sheave is mounted on the high speed shaft, such as the shaft of an electric motor. In general, belt drives are applied where the rotational speeds are relatively high, as on the first stage of speed reduction from an electric motor or engine. The linear speed of a belt is usually 2,500 to 7,000 ft/min which results in relatively low tensile forces in the belt. At lower speeds, the tension in the belt becomes too large for typical belt cross sections, and slipping may occur between the sides of the belt and the sheave or pulley that carries it. At higher speeds, dynamic effects such as centrifugal forces, belt whip, and vibration reduce the effectiveness of the drive and its life. A speed of 4,000 ft/min is generally ideal. Some belt designs employ high strength, reinforcing strands and a cogged design that engages matching grooves in the pulleys to enhance their ability to transmit the high forces at low speeds. These designs compete with chain drives in many applications. Sheave is mounted on the driven machine. The belt is designed to ride around the two sheaves without slipping.

The belt is installed by placing it around the two sheaves while the center distance between them is reduced. Then the sheaves are moved apart, placing the belt in a rather high initial tension. When the belt is transmitting power, friction causes the belt to grip the driving sheave, increasing the tension in one side, called the "tight side" of the drive. The tensile force in the belt exerts a tangential force on the driven sheave, and thus a torque is applied to the driven shaft. The opposite side of the belt is still under tension, but at a smaller value. Thus, it is called the "slack side".

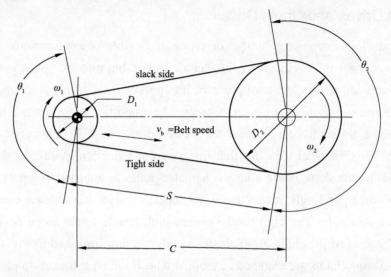

Figure 4.6　Basic belt drive schematic geometry

Many types of belts are available: flat belts, grooved or cogged belts, standard V-belts, double-angle V-belts, etc.

The most widely used type of belt, particularly in industrial drives and vehicular applications, is the V-belt drive, as shown in Figure 4.7. The V-shape causes the belt to wedge tightly into groove, increasing friction and allowing high torques to be transmitted before slipping occurs. Most belts have high strength cords positioned at the pitch diameter of the belt cross section to increase the tensile strength of the belt. The cords, made from natural fibers, synthetic strands, or steel, are embedded in a firm rubber compound to provide the flexibility needed to allow the belt to pass around the sheave. Often an outer fabric cover is added to give the belt good durability.

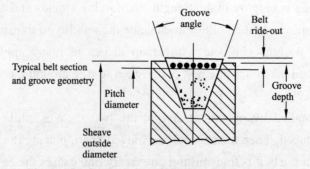

Figure 4.7　Cross section of V-belt and sheave groove

Many design decisions depend on the application and on space limitations. A few guidelines are given as follows:

- Adjustment for the center distance must be provided in both directions from the nominal value. The center distance must be shortened at the time of installation to enable the belt to be placed in the grooves of the sheaves without force. Provision for increasing the center distance must be made to permit the initial tensioning of the drive and to take up for belt stretch.
- If fixed centers are required, idler pulleys should be used. It is best to use a grooved idler on the inside of the belt, close to the sheave.
- The nominal range of center distance should be
$$0.7(D_1 + D_2) < C < 2(D_1 + D_2) \tag{4.3}$$
- The angle of wrap on the smaller sheave should be greater than 120°.
- Ensure that the shafts carrying mating sheaves are parallel and that the sheaves are in alignment so that the belts track smoothly into the grooves.
- Belts must be installed with the initial tension recommended by the manufacturer. Tension should be checked after the first hours of operation because seating and initial stretch occur.

A chain is a power transmission element made as a series of pin-connected links. The design provides for flexibility while enabling the chain to transmit large tensile forces. When transmitting power between rotating shafts, the chain engages mating toothed wheels, called sprockets.

The most common type chain is the roller chain, in which the roller on each pin provides exceptionally low friction between the chain and the sprockets. Roller chain is classified by its pitch, the distance between corresponding parts of adjacent links. The pitch is usually illustrated as the distance between the centers of adjacent pins.

The rating of chain for its power transmission capacity considers three modes of failure:

- Fatigue of the link plates due to the repeated application of the tension in the tight side of the chain;
- Impact of the rollers as they engage the sprocket teeth;
- Galling between the pins of each link and the bushing on the pins.

4.2.2 Gears and Wormgear Drives

Gears are toothed, cylindrical wheels used for transmitting motion and power from one rotating shaft to another. The teeth of a driving gear mesh accurately in the spaces between teeth on the driven gears as shown in Figure 4.8. The driving teeth push on the driven teeth, exerting a force perpendicular to the radius of the gear. Thus, a torque is transmitted, and because the gear is rotating, power is also transmitted.

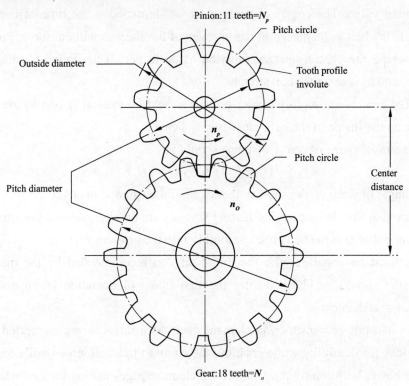

Figure 4.8 Pair of spur gears

Often gears are employed to produce a change in the speed of rotation of the driven gear relative to the driving gear. In Figure 4.8, if the smaller top gear, called a pinion, is driving the larger lower gear, simply called the gear, the larger gear will rotate more slowly. The amount of speed reduction is dependent on the ratio of the number of teeth in the pinion to the number of teeth in the gear according to the relationship as follows:

$$n_P/n_G = N_G/N_P \tag{4.4}$$

Spur gears have teeth that are straight and arranged parallel to the axis of the shaft that carries the gear. The curved shape of the faces of the spur gear teeth have a special geometry called an involute curve. This shape makes it possible for two gears to operate together with smooth, positive transmission of power.

The most widely used spur gear tooth form is the full-depth involute form. The involute is one of a class of geometric curves called conjugate curves. When two such gear teeth are in mesh and rotating, there is a constant angular velocity ratio between them. From the moment of initial contact to the moment of disengagement, the speed of the driving gear is in a constant proportion to the speed of the driven gear. The resulting action of the two gears is very smooth. If this were not the case, there would be some speeding up and slowing down during the engagement, with the resulting accelerations causing vibration, noise, and dangerous torsional oscillations in the system.

The gears and the wormgear drives

Figure 4.9 shows drawings of spur gear teeth, with the symbols for the various features indicated.

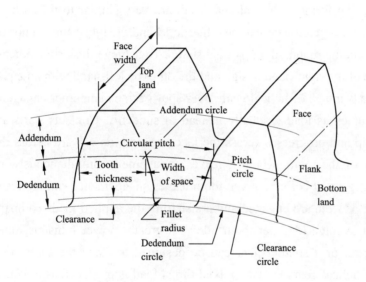

Figure 4.9　Spur gear teeth features

When two gears are in mesh, there are two circles. One from each gear, that remain tangent throughout the engagement cycle. These are called the pitch circles. The diameter of the pitch circle of a gear is its pitch diameter; the point of tangency is the pitch point.

When two gears mesh, the smaller gear is called the pinion, and the larger is the gear. The spacing between adjacent teeth and the size of the teeth are controlled by the pitch of the teeth. The pitch of two gears in mesh must be identical. The pitch of gears in the metric system is designated the module, m, and it is attained by dividing the pitch diameter of the gear in millimeters by the number of teeth.

The pressure angle is the angle between the tangent to the pitch circles and the line drawn normal (perpendicular) to the surface of the gear tooth. Standard values of the pressure angle are established by gear manufactures (usually is 20°), and the pressure angles of two gears in mesh must be the same.

The wormgear drives

For certain combinations of numbers of teeth in a gear pair, there is interference between the tip of the teeth on the pinion and the fillet or root of the teeth on the gear. Obviously this cannot be tolerated because the gears simply will not mesh. The probability that interference will occur is greatest when a small pinion drives a large gear, with the worst case being a small pinion driving a rack. A rack is a gear with a straight pitch line; it can be thought of as a gear with an infinite pitch diameter. To avoid the interference, the surest way is to control the minimum number of teeth in the pinion to the limiting values.

On helical gears, the teeth are inclined at an angle with the axis, and that angle is called the helix angle. The forms of helical gear teeth are very similar to those discussed for spur gears. The helix for a given gear can be either left-hand or right-hand. In normal installation, helical gears would be mounted on parallel shafts. To achieve this arrangement, it is required that one gear be of the right-hand design and the other be left-hand with an equal helix angle.

The main advantage of helical gears over spur gears is smoother engagement because a given tooth assumes its load gradually instead of suddenly. Contacts starts at one end of a tooth near the tip and progresses across the face in a path downward across the pitch line to the lower flank of the tooth, where it leaves engagement. Simultaneously, other teeth are coming into engagement before a given tooth leaves engagement, with the result that a larger average number of teeth are engaged and are sharing the applied loads compared with a spur gear. The lower average load per tooth allows a greater power transmission capacity for a given size of gear, or a smaller gear can be designed to carry the same power. The main disadvantage of helical gears is that an axial thrust load is produced as a natural result of the inclined arrangement of the teeth. The bearing that hold the shaft carrying the helical gear must be capable of reacting against the thrust load.

Bevel gears are used to transfer motion between nonparallel shafts, usually at 90° to one another. The teeth of a straight bevel gear are straight and lie along an element of the conical surface. Lines along the face of the teeth through the pitch circle meet at the apex of the pitch cone. The centerlines of both the pinion and the gear also meet at this apex. In the standard configuration, the teeth are tapered toward the center of the cone. The mounting of bevel gears is critical if satisfactory performance is to be achieved.

Wormgearing is used to transmit motion and power between nonintersecting shafts, usually at 90° to each other. The drive consists of a worm on the high speed shaft which has the general appearance of a power screw thread: a cylindrical, helical thread. The worm drives a wormgear, which has an appearance similar to that of a helical gear. Sometimes the wormgear is referred to as a worm wheel or simply a wheel or gear. Several variations of the geometry of wormgear drives are available. The most common one, employs a cylindrical worm mating with a wormgear having teeth that are throated, wrapping partially around the worm. The contact between the threads of the worm and wormgear teeth is along a line, and the power transmission capacity is quite good.

4.2.3 Keys and Couplings

Think of how two or more parts of a machine can be connected for the purpose of locating one part with respect to another. Now think about how that connection must be designed if the parts are moving and if power must be transmitted between them.

A key is used to connect a drive member such as a belt pulley, chain sprocket, or gear to the shaft that carries it. Torque and power are transmitted across the key to or from the shaft. But how does the power get into or out of the shaft? One way might be that the output from the shaft of a motor or engine is connected to the input shaft of a transmission through a flexible coupling that reliably transmits power but allows for some misalignment between the shafts during operation because of the flexing of frame members or through progressive misalignment due to wear.

The keys and couplings

A key is a machinery component placed at the interface between a shaft and the hub of a power-transmitting element for the purpose of transmitting torque. The key is demountable to facilitate assembly and disassembly of the shaft system. It is installed in an axial groove machined into the shaft, called a keyseat. A similar groove in the hub of the power-transmitting element is usually called a keyway, but it is more properly also a keyseat. The key is typically installed into the shaft keyseat first; then the hub keyseat is aligned with the key, and the hub is slid into position.

The most common types of key for shafts is parallel key because the top and bottom and the sides of the key are parallel. The keyseats in the shaft and the hub are designed so that exactly one-half of the height of the key is bearing on the side of the shaft keyseat and the other half on the side of the hub keyseat. Taper keys are designed to be inserted from the end of the shaft after the hub is in position rather than installing the key first and then sliding the hub over the key as with parallel keys. The taper extends over at least the length of the hub, and the height, measured at the end of the hub, is the same as for the parallel key. The gib head key has a tapered geometry inside the hub that is the same as that of the plain taper key. But the extended head provides the means of extracting the key from the same end at which it was installed. This is very desirable if the opposite end is not accessible to drive the key out.

The key and the keyseat for a particular application are usually designed after the shaft diameter is specified. Then, with the shaft diameter as a guide, the size of the key is selected. The only remaining variables are the length of the key and its material. One of these can be specified, and the requirements for the other can be computed.

The term coupling refers to a device used to connect two shafts together at their ends for the purpose of transmitting power. There are two general types of couplings: rigid and flexible.

Rigid couplings are designed to draw two shafts together tightly so that no relative motion can occur between them. This design is desirable for certain kinds of equipment in which precise alignment of two shafts is required and can be provided. In such cases, the coupling must be designed to be capable of transmitting the torque in the shaft. Rigid couplings should be used only when the alignment of the two shafts can be maintained very accurately, not only at the time of installation but also during operation of the machine. If significant angular, raidal, or axial misalignment occurs, stresses that are difficult to predict and that may lead to early failure due to fatigue will be induced in the shafts. These difficulties can be overcome by the use of flexible couplings.

Flexible couplings are designed to transmit torque smoothly while permitting some axial, radial, and angular misalignment. The flexibility is such that when misalignment does occur, parts of the coupling move with little or no resistance. Thus, no significant axial or bending stresses are developed in the shaft. Many types of flexible couplings are available commercially. Each is designed to transmit a given limiting torque. The manufacture's catalog lists the design data from which you can choose a suitable coupling. Remember that torque equals power divided by rotational speed, so for a given size of coupling, as the speed of rotation increases, the amount of power that the coupling can transmit also increases, although not always in direct proportion. Of course, centrifugal effects determine the upper limit of speed.

4.2.4 Bearings

The bearings

The purpose of a bearing is to support a load while permitting relative motion between two elements of a machine. The term rolling contact bearings refer to the wide variety of bearings that use spherical balls or some other type of roller between the stationary and the moving elements. The most common type of bearing supports a rotating shaft, resisting purely radial loads or a combination of radial and axial (thrust) loads. Some bearings are designed to carry only thrust loads. Most bearings are used in applications involving rotation, but some are used in linear motion applications. The term plain surface bearing refers to the kind of bearing in which two surfaces move relative to each other without the benefit of rolling contact. Thus, there is sliding contact. The actual shape of the surfaces can be anything that permits the relative motion. The most common shapes are flat surfaces and concentric cylinders.

(1) Rolling contact bearings

The components of a typical rolling contact bearing are the inner race, the outer race, and the rolling elements. Usually the outer race is stationary and is held by the housing of the machine. The inner race is pressed onto the rotating shaft and thus rotates with it. Then the balls roll between the outer and inner races. The load path is from the shaft to the inner race, to the balls, to the outer race, and finally to the housing. The presence of the balls allows a very smooth, low-friction rotation of the shaft. The typical coefficient of friction for a rolling contact bearing is approximately 0.001 to 0.005. These values reflect only the rolling elements themselves and the means of retaining them in the bearing. The presence of seals, excessive lubricant, or unusual loading increases these values.

Here we will discuss six different types of rolling contact bearings and the applications in which each is typically used. Table 4.2 provides a comparison of the performance relative to the others.

Table 4.2 Comparison of bearing types

Bearing type	Radial load capacity	Thrust load capacity	Misalignment capacity
Single-row, deep-groove ball	Good	Fair	Fair
Angular contact	Good	Good	Fair
Cylindrical roller	Excellent	Poor	Fair
Needle	Excellent	Poor	Poor
Spherical roller	Excellent	Fair/good	Excellent
Tapered roller	Excellent	Excellent	Poor

Radial loads act toward the center of the bearing along a radius. Such loads are typical of those created by power transmission elements on shafts such as spur gears, V-belt drives, and chain drives. Thrust loads are those that act parallel to the axis of the shaft. The axial components of the forces on helical gears, worms and wormgears, and bevel gears are thrust loads. Also, bearings supporting shafts with vertical axes are subjected to thrust loads due to the weight of the shaft and the elements on the shaft as well as from axial operating forces. Misalignment refers to the angular deviation of the axis of the shaft at the bearing from the true axis of the bearing itself.

The single-row, deep-groove ball bearing is sometimes called Conrad bearings. The inner race is typically pressed on the shaft at the bearing seat with a slight interference fit to ensure that it rotates with the shaft. The spacing of the balls is maintained by retainers or "cages". While designed primarily for radial load-carrying capacity, the deep groove allows a fairly sizable thrust load to be carried. Because the load is carried on a small area, very high local contact stresses occur.

One side of each race in an angular contact bearing is higher to allow the accommodation of greater thrust loads compared with the standard single-row, deep-groove bearing. The preferred angle of the resultant forc (radial and thrust loads combined) is of 15° to 40°.

Cylindrical roller bearing is replaced the spherical balls with cylindrical rollers, with corresponding changes in the design of the race. It gives a greater radial load capacity. The resulting contact stress levels are lower than for equivalent-sized ball bearings, allowing smaller bearings to carry a given load or a given size bearing to carry a higher load. Thrust load capacity is poor because any thrust load would be applied to the side of the rollers, causing rubbing, not true rolling motion.

Needle bearings are actually roller bearings, but have much smaller-diameter rollers. A smaller radial space is typically required for needle bearings to carry a given load than for any other type of rolling contact bearing.

The spherical roller bearing is one form of self-aligning bearing, so called because there is actual relative rotation of the outer race relative to the rollers and the inner race when angular misalignments occur.

Tapered roller bearings are designed to take substantial thrust loads along with high radial loads, resulting in excellent ratings on both. They are often used in wheel bearings for vehicles and mobile equipment in heavy-duty machinery having inherently high thrust loads.

The load on a rolling contact bearing is exerted on a small area. The resulting contact stresses are quite high, regardless of the type of bearing. To withstand high stresses, the balls, rollers, and races are made from a very hard, high strength steel or ceramic.

Despite using steels with very high strength, all bearings have a finite life and will eventually fail due to fatigue because of the high contact stresses. But, obviously, the lighter the load, the longer the life, and vice versa. The relationship between load, P, and life, L, for rolling contact bearings can be stated as

$$\frac{L_2}{L_1} = \left(\frac{P_1}{P_2}\right)^k \tag{4.5}$$

where $k=3.0$ for ball bearings and $k=3.33$ for roller bearings.

(2) Journal bearings

The name journal bearing is sometimes used for plain surface bearings. This term is derived from the terminology of the components of the complete bearing system. For the case of a bearing shaft, the portion of the rotating shaft at the bearing is called the journal. The stationary part that supports the load is the bearing. Figure 4.10 shows the basic geometry of a journal bearing. A given bearing system can operate with any of three types of lubrication: boundary lubrication, mixed-film lubrication, and full-film lubrication. All of these types of lubrication can be encountered in a bearing without external pressurization of the bearing.

The performance of a bearing differs radically, depending on which type of lubrication occurs [1]. There is a marked decrease in the coefficient of friction when operation changes from boundary to full-film lubrication. Wear also decreases with full-film lubrication. Thus, it is desirable to understand the conditions under which one or the other type of lubrication occurs. The creation of full-film lubrication, the most desirable type, is encouraged by light loads, high relative speed between the moving and stationary parts, and presence of a high viscosity lubricant in the bearing in copious supply. For a rotating journal bearing, the combined effect of these three factors, as it relates to the friction in the bearing, can be evaluated by computing the bearing number, $\eta n/p$. The viscosity of the lubricant is indicated by η, the rotational speed by n and the bearing load by the pressure p.

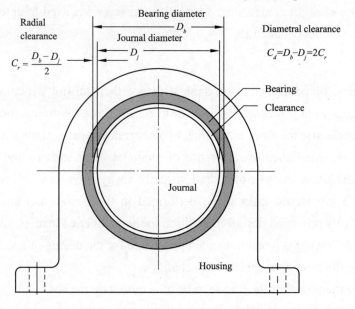

Figure 4.10 Journal bearing geometry

In rotational applications, the journal on the shaft is frequently steel. The stationary bearing may be made of any of a wide variety of materials, including the following: bronze, Babbitt, aluminum, porous metals, and plastics (nylon, TFE, PTFE, phenolic, acetal, polycarbonate, filled polyimide, etc). The properties desirable for materials used for plain bearings are unique, and compromised must frequently be made. These properties include, among some others: strength, embeddability, corrosion resistance, cost.

4.2.5 Fasteners

A fastener is any device used to connect or join two or more components. Literally hundreds of fastener types and variations are available. The most common are threaded fasteners referred to by many names, among them bolts, screws, nuts, studs, lag screws, and set screws.

A bolt is a threaded fastener designed to pass through holes in the mating members and to be secured by tightening a nut from the end opposite the head of the bolt. A screw is a threaded fastener designed to be inserted through a hole in one member to be joined and into a threaded hole in the mating member. Machine screws, also called cap screws, are precision fasteners with straight-threaded bodies that are turned into tapped holes. Sheet-metal screws, lag screws, self-tapping screws, and wood screws usually form their own threads.

In machine design, most fasteners are made from steel because of its high strength, good ductility, and good machinability and formability. But varying compositions and conditions of steel are used. The strength of steels used for bolts and screws is used to determine its grade, according to one of several standards.

4.2.6 Shaft

The shaft

A shaft is the component of a mechanical device that transmits rotational motion and power. It is integral to any mechanical system in which power is transmitted from a prime mover, such as an electric motor or an engine, to other rotating parts of the system.

Because of the simultaneous occurrence of torsional shear stresses and normal stresses due to bending, the stress analysis of a shaft virtually always involves the use of a combined stress approach. The specific tasks to be performed in the design and analysis of a shaft depend on the shaft's proposed design in addition to the manner of loading and support. With this in mind, the following is a recommended procedure for the design of a shaft.

- Determine the rotational speed of the shaft.
- Determine the power or the torque to be transmitted by the shaft.
- Determine the design of the power-transmitting components or other devices that will be mounted on the shaft, and specify the required location of each device.
- Specify the location of bearings to support the shaft.
- Propose the general form of the geometry for the shaft, considering how each element on the shaft will be held in position axially and how power transmission from each element to the shaft is to take place.
- Determine the magnitude of torque that the shaft sees at all points. It is recommended that a torque diagram be prepared.
- Determine the forces that are exerted on the shaft, both radially and axially.

- Resolve the radial forces into components in perpendicular directions, usually vertically and horizontally.
- Solve for the reactions on all support bearings in each plane.
- Produce the complete shearing force and bending moment diagrams to determine the distribution of bending moments in the shaft.
- Select the material from which the shaft will be made, and specify its condition.
- Determine an appropriate design stress, considering the manner of loading.
- Analyze each critical point of the shaft to determine the minimum acceptable diameter of the shaft at that point in order to ensure safety under the loading at that point.
- Specify the final dimensions for each point on the shaft.

Words and Expressions

journal bearing [ˈdʒɜːnl ˈbeərɪŋ]	径向轴承
viscosity [vɪˈskɒsəti] resistance of a liquid to sheer forces (and hence to flow)	*n.* 黏度

Notes

[1] 本句可译为：所处的润滑状态不同，轴承将具有完全不同的润滑行为。

Questions for Discussion

Decompose a typical machine to make acquaintance with the machine elements commonly found in mechanical devices.

4.3 Friction, Wear and Lubrication

By definition, tribology is the science and technology of interacting surfaces in relative motion (or having a propensity to move relatively) and of related subjects and practices. In literature, it includes friction, wear, and lubrication for the most time. These three processes affect each other with interacting causes and effects, making the issue a complex problem. Interactions between surfaces in a tribological interface are highly complex, and their fully understanding requires knowledge of various disciplines including physics, chemistry, applied mathematics, solid mechanics, fluid mechanics, thermodynamics, heat transfer, materials science, rheology, lubrication, machine design, performance, reliability, and etc.

Friction, wear and lubrication

It was the artist Leonardo da Vinci who first gave a scientific approach to friction process. He introduced, for the first time, the concept of coefficient of friction (COF) as the ratio of the friction force to the normal load and the rules of friction are later rediscovered by Amontons and verified by Coulomb. Lubrication theories are established by Reynolds after experimental studies of Beauchamp Tower, showing that a continuously film can separate journal surfaces and the film being maintained by the motion of the journal. Wear is a much younger subject than friction and lubrication, and it was initiated on empirical basis to a large extent. The purpose of research in tribology is usually understandably the minimization and elimination of losses resulting from friction and wear at all levels of technology where the rubbing of surfaces is involved. However, to the other side of the same coin, wear is adopted as a manufacturing method to acquire a surface with high precision, such as in a chemical mechanical polishing. Or in some other situations, for example, in a MEMS system, wear is a useful process to acquire a surface with optimized surface topography.

4.3.1 Friction

Friction

Friction is the force arising to resist relative motion (sliding or rolling) that is experienced when one solid body moves tangentially over another with which it is in intimate contact. Friction will give rise to retardation of relative motion and, to be more scientifically, a loss of energy of the system. Friction is not a material property; it is a system response. In a contact of solid sliding on solid, the rules of friction are broadly adopted as follows:

- Firstly, the friction force is directly proportional to the exerted normal load.
- Secondly, the amount of friction force (and thereby the COF) does not depend upon the apparent area of contact.
- Thirdly, the friction force is independent of the velocity once motion starts.
- And the fourth, static friction is always larger than kinetic friction.

On conditions cited above, the relation between the friction force (F) and the load (W) obeys the following equation:

$$F = \mu W \tag{4.6}$$

where μ is a constant known as COF.

When two nominally flat surfaces are placed in intimate contact under exerted load, the "real" contact only takes place at the tips of the asperities, the load being supported by the deformation of contacting asperities, either elastically or plastically, and discrete contact spots are formed. Bowden and Tabor proposed that for two sliding metals, high pressures developed at individual contact spots will lead to local welding and the contacts thus formed are sheared subsequently by relative sliding of the surfaces. Later, they argued that asperities do not have to weld, but that just interfacial adhesion between asperities is sufficient enough to account for friction of ceramics. The dominant mechanism of energy dissipation in ceramics is assumed to be plastic deformation. If one assumes that the interaction between the adhesion and deformation processes during sliding is negligible, the total intrinsic frictional force (F_i) equals the force needed to shear adhered junctions (F_a), regardless of the type of deformation (either elastic or plastic), breaking of adhesive bonds during motion need energy dissipation, and the force needed to supply the energy of deformation (F_d).

$$F_i = F_a + F_d \tag{4.7}$$

or the coefficient of friction $\mu_i = \mu_a + \mu_d$. The distinction between the adhesion and deformation theories is arbitrary.

Adhesion coefficient of friction μ_a, is determined as follows:

$$\mu_a = \frac{\tau_a}{H}\left(1 + K\frac{W_{ad}}{H}\right) \tag{4.8}$$

where τ_a is the average shear strength of the contact; H is the hardness of the softer of the contacting materials; K is a geometric factor; W_{ad} is work of adhesion.

During any relative motion in practice, adhesion and asperity interactions are always present, though the ploughing contribution may or may not be significant. Ploughing not only increases friction force, it also enriches wear particles, which in turn increase subsequent friction and wear. In the case of ceramic pairs with two rough surfaces or with trapped hard particles, the deformation term contributes to the force needed for ploughing, grooving or cracking surfaces, and it is generally dominant compared to the adhesion component. The ploughing component of the friction force should be determined according to the geometric model of the particles. In addition, for brittle materials (such as ceramics), fracture of adhesive contacts and brittle deformation of materials need to be considered. An additional material property, i.e., fracture toughness is important.

4.3.2　Wear

Wear is the surface damage or removal of material from one or both of the contacting surfaces in a system of relative motion, or the change in surface topography. In most cases, wear occurs through surface interactions at asperities. Definition of wear is generally based on loss or transfer of material, but it should be emphasized that change of surface condition due to material displacement on a given body, with no net change in weight or volume, also constitutes wear.

Wear

Wear, as friction, is not a material property but a system response. Wear is closely related to friction, and it is sometimes assumed that high friction will always result in high wear rates. This is not necessarily true. Similarly, wear can be either good or not. For most industrial applications, however, wear is undesirable.

(1) Types of wear mechanisms

1) Adhesive wear

Adhesive wear occurs when two nominally flat solid bodies are in sliding contact, whether lubricated or not. Adhesion occurs at the asperity contacts at the interface, and these contacts are sheared by relative motion, which may result in detachment of a fragment from one surface and attachment to the other surface. As the sliding continues, the transferred fragments may come off the surface on which they are transferred and be transferred back to the original surface, or else form loose wear particles. Some are fractured by a fatigue process during repeated loading and unloading actions.

Archard pointed out that shearing can occur at the original interface or in the weakest region in one of the two bodies as shown in Figure 4.11. In most cases, the break during shearing process occurs at the interface (path 1) in most of the contacts and no wear occurs in that sliding cycle. In a small fraction of contacts, break may take place in one of the two bodies (path 2) and a small fragment may attach to the other mating surface. Plastic shearing of successive layers can also result in detachment of a wear fragment. The fragment is detached from one surface and then transferred to the mating surface because of adhesion.

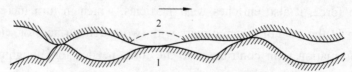

Figure 4.11 Schematic view showing two possibilities of break (1 and 2) during shearing

The amount of wear (material loss) is considered generally proportional to the applied load W and sliding distance x, and generally inversely proportional to the hardness H of the surface being worn away. That is, the volume of wear being worn away is (Holm)

$$v = k\frac{W}{H}x \tag{4.9}$$

where k is a non-dimensional wear coefficient related to the material in contact.

2) Abrasive wear

When asperities of a rough, hard surface or hard particles slide on a softer surface and damage the interface by plastic deformation or fracture, abrasive wear occurs. There are two general situations for abrasive wear. In one case, the hard surface is the harder of the two rubbing surfaces (also known as two-body abrasion). In the other case, the hard surface is a third body (also known as three-body abrasion). In many cases, the wear mechanism at the start is adhesive, which generates wear particles that get trapped at the interface, resulting in a three-body abrasive wear.

a) Abrasive wear by plastic deformation

Material removal from a surface via plastic deformation can occur by several deformation modes which include ploughing, wedge formation and cutting. To obtain a quantitative expression for abrasive wear for plastic contacts, a simple model is considered, in which one surface consists of an array of hard conical asperities sliding on a softer and flat surface and ploughs a groove of uniform depth. An equation of the form similar to Archard's equation for adhesive wear is found to cover a wide range of abrasive situations as

$$v = k_{abr} \frac{W}{H} x \tag{4.10}$$

where k_{abr} is a non-dimensional wear coefficient that includes the geometry of the asperities and the probability that given asperities cut rather than plough.

b) Abrasive wear by fracture

At low loads, a sharp asperity contact will cause only plastic deformation and wear occurs by plastic deformation. Above a threshold load, brittle fracture occurs, and wear occurs by lateral cracking at a sharply increased rate. The threshold load is proportional to $(K_c/H)^3 K_c$. The H/K_c is known as index of brittleness, where K_c is fracture toughness.

The volume of wear per unit sliding distance of the interface is given by

$$v = \alpha N \frac{(E/H) W^{9/8}}{K_c^{1/2} H^{5/8}} \tag{4.11}$$

where E is the Young's model; α is a material independent constant; N is the number of asperities that contact the surface.

3) Fatigue wear

Fatigue involves the cyclically repeated stresses on the contact region until permanent damage occurs. It is usually assumed that since the internal structure of ordinary materials is not perfect, grain boundaries, inclusions, voids, and the like can act as stress concentration sites to initiate formation of cracks which propagate with repeated stressing until a spalled particle detaches from the body of the material. But even if the substance is nearly perfect crystallographically, it is theoretically still possible for dislocations to appear when the material deforms under the first stressing and then to grow under subsequent cycles of stress.

4) Chemical (corrosive) wear

Chemical or corrosive wear occurs when rubbing takes place in a corrosive environment. Chemical wear in air is generally called oxidative wear, due to the fact that the most dominant corrosive medium to metals in air is oxygen. Chemical wear requires both chemical reaction and rubbing (by sliding). Corrosion can occur because of chemical of electrochemical interaction of the surface with the environment.

The purely chemical and the tribochemical aspects of the chemical reaction process are distinguishable. In the purely chemical case, the material of the surface is capable of reacting directly with the constituents in the ambient environment (oxygen or water) under the ordinarily prevailing ambient conditions; but for tribochemical action the surface must be activated by rubbing process for reaction to occur. We can draw a distinction between the case of the tribologically activated surface reaction competing directly with the tribological removal of the reaction products and the case of ordinary chemical reaction with environmental constituents followed by a separate rubbing process that removes the reaction products.

5) Fretting and fretting corrosion

Fretting wear occurs between two surfaces in nominal stationary but really with oscillatory relative motion of small amplitude. In principle, fretting is a form of adhesive or abrasive wear, or the combination of the two. The normal load causes adhesion between asperities and oscillatory movement causes ruptures resulting in wear debris. Most commonly, fretting is accompanied with corrosion, heralding fretting corrosion wear. To minimize fretting wear, the machinery should be designed to reduce oscillatory movement, reduce stresses or eliminate two-piece design altogether.

(2) Types of particles in wear debris

The size and shape of wear debris may change during sliding. Mild wear is characterized by finely divided wear debris and the worn surface is relatively smooth. Severe wear, in contrast, results in much larger particles, which may be even visible with the naked eye, and the worn surface is rough. Particles can be classified based on wear mechanism or their morphology.

a) Plate-shaped particles

These particles are produced as a result of ploughing followed by repeated loading and unloading fatigue, resulting from nucleation and propagation of subsurface cracks or plastic shear in the asperity contacts. Thin, plate-shaped or flake-type wear particles commonly have an aspect ratio of 2 to 10.

b) Ribbon-shaped particles

Ribbon-shaped particles are produced as a result of plastic deformation. They are usually curved and even curly with aspect ratios on the order of ten or more. These particles are produced with sharp asperities or abrasive particles (usually two-body abrasive wear) digging into the mated surface with materials flowing up the front face of the asperity or abrasive particles and being detached from the wearing surface in the form of a chip.

c) Spherical particles

Wear particles of various shapes may not escape from the interface to become loose debris. Some of them remain trapped and are processed further as in the spherical shape.

d) Irregular-shaped particles

The majority of wear particles have irregular morphology. Wear debris produced by detachment of the transferred fragment in adhesive wear and brittle fracture are always irregular.

4.3.3 Lubrication

Lubrication is usually applied to a tribo system to mitigate friction and wear. Lubrication is the process, or technique employed to reduce wear of one or both surfaces in close proximity, and moving relative to each another, by interposing a substance called lubricant to provide reduced shear strength between the surfaces to carry or to help carry the load. There are regimes of lubrication in practice, depending upon the conditions in practices.

Lubrication

More than a century ago, Stribeck represented the variation of the friction coefficient (μ) for a sliding bearing. In a Stribeck curve, for a hypothetical fluid-lubricated bearing system, the COF is presented as a function of the product of the absolute viscosity (η) and rotational speed in revolutions per unit second (N) divided by the pressure (p). The regimes of lubrication are sometimes identified by a lubricant film parameter (λ) equal to h/σ (mean film thickness / composite standard deviation of surface heights of the two surfaces). The composite roughness is defined as

$$\sigma = \sqrt{\sigma_1^2 + \sigma_2^2} \tag{4.12}$$

where σ_1 and σ_2 are surface roughness of the two contacting bodies, respectively. By using the lubricant film parameter, one can also judge the lubrication regime, i.e. the lubrication regime is elastohydrodynamic lubrication (EHL) or hydrodynamic lubrication (HL) if $\lambda \geqslant 3$, mixed lubrication if λ ranges from 1 to 3, and boundary lubrication (BL) when λ is smaller than 1.

The method abovementioned is not applicable for the ultra-flat solid surfaces. Therefore, the distinguishing method for different lubrication regimes should not only include the lubricant film parameter, but also the number of the molecular layers, or the ratio of film thickness to lubricant molecular size (λ'). Luo et al proposed a new method to distinguish the lubrication regimes as shown in Figure 4.12. When λ is larger than 3, three lubrication regimes can be formed separately according to λ'. The regime is boundary lubrication when λ' is

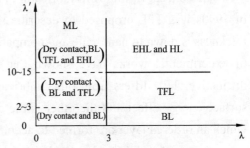
Figure 4.12 Lubrication map

smaller than about 3, is thin film lubrication (TFL) when λ' is larger than 3 and lower than 10 to 15 which is related to the surface force of the solid surfaces and molecular polarity, and is EHL or HL when the ratio λ' is higher than 15. The mixed lubrication will be found when the ratio λ is smaller than 3.

where ML: mixed lubrication; EHL: elastohydrodynamic lubrication; TFL: thin film lubrication; BL: boundary lubrication, HL: hydrodynamic lubrication.

(1) Hydrodynamic lubrication (HL)

Hydrodynamic lubrication (HL) is sometimes called full film or thick film lubrication. At the beginning of the 20th century, HL was well advanced, based on the work of Navier and Stokes. Reynolds' equation had been widely validated for continuous liquid films acting only under restricted conditions. The fluid film is thick enough to prevent physical / direct contact of the friction couple elements. Within HL, the couple elements may function without any measurable wear for a long operating time.

(2) Elastohydrodynamic lubrication (EHL)

Elastohydrodynamic lubrication (EHL) is the typical regime for friction pairs with elastic contact under very high pressure in an unconformal contact such as ball bearings and gears[1]. EHL is a development of hydrodynamic lubrication to take into account the elastic deflection of the solid surfaces in contact. Film thicknesses are much smaller than in conventional HL. Under these conditions, a continuous hydrodynamic film can only be achieved through elastic deflections (usually in combination with an increase in viscosity due to the high contact pressure).

The theory of the EHL regime was established in the latter half of the 20th century, notably by Dowson and Higginson and further by Winer and Cheng. Due to very high pressures at the contact interface, the EHL regime also involves elastic smoothing of the surface micro-asperities. By this mechanism and by an increase in viscosity due to extreme pressure, EHL ensures continuity of the fluid film in a bearing contact.

(3) Thin film lubrication (TFL)

The concept of thin film lubrication (TFL) is relatively new. Only until at the beginning of 1990s was TFL proposed to describe the lubrication regime in which the lubricant film thickness is down to nanoscale and comparable to the molecular size of the lubricant. A series of experimental works have shown that the film forming properties in TFL regime are distinctive. TFL differs from other lubrication regimes in two main contributors: one is the surface energy effect and the other is the microstructure of the lubricant molecules, through which an ordered layer are formed to account for the lubrication.

(4) Boundary lubrication (BL)

Boundary lubrication is that condition in which the solid surfaces are so close together that surface interaction between monomolecular or multi-molecular films of lubricants (liquids or gases) and the solid asperities dominates the contact. The failure in boundary lubrication occurs by adhesive and chemical (corrosive) wear. Boundary lubricants form an easily sheared film on the bearing surfaces, thereby minimizing adhesive wear and chemical wear. The important physical properties of the films are the melting point, shear strength and hardness. Other properties are adhesion or tenacity, cohesion and rates of formation. The bulk flow properties of the lubricant (such as viscosity) play little part in the friction and wear behavior.

(5) Mixed lubrication (ML)

Mixed lubrication is a regime in which two or more lubrication mechanisms may be functioning spontaneously. There may be more frequent solid contacts, but at least a portion of the bearing surface remains supported by a partial hydrodynamic fluid film. The solid contacts, between unprotected virgin solid surfaces, could lead to a cycle of adhesion, metal transfer, wear particle formation and detachment, and eventual seizure. However, in liquid lubricated bearings, physico- or chemo-sorbed or chemically reacted films can prevent adhesion from most asperity encounters.

Words and Expressions

tribochemical [ˈtraɪbəʊ ˈkemɪkəl]	摩擦化学
conformal contact [kɔnˈfɔːml ˈkɔntækt]	共形接触

Notes

[1] 本句可译为：弹性流体动力润滑(EHL)是球轴承、齿轮等非共形接触中超高压弹性接触摩擦副的典型润滑形式。

Questions for Discussion

Why are the friction coefficient and the wear coefficient related to the friction process other than the materials involved wherein?

4.4 Industrial Robot

4.4.1 Introduction

In the study of robotics, the connection between the field of study and ourselves is unusually obvious. And, unlike a science that seeks only to analyze, robotics are presently pursued takes the engineering bent toward synthesis. The study of robotics concerns itself with the desire to synthesize some aspects of human function by the use of mechanisms, sensors, actuators, and computers. Presently different aspects of robotics research are carried out by experts in various fields. At a relatively high level of abstraction, splitting robotics into four major areas seems reasonable: mechanical manipulation, locomotion, computer vision, and artificial intelligence. In this section, industrial robots are adopted to give brief elucidation of this fascinating area.

Industrial robots are relatively new electromechanical devices that are beginning to change the appearance of modern industry. Industrial robots are not like the science fiction devices that possess human-like abilities and provide companionship with space travelers. Research to enable robots to "see", "hear", "touch", and "listen" has been underway for two decades and is beginning to bear fruit. However, the current technology of industrial robots is such that most robots contain only on arm rather than all the anatomy a human possess. Current control only allows these devices to move from point to point in space, performing relatively simple tasks. The Robotics Institute of America defines a robot as "a reprogrammable multifunction manipulator designed to move material, parts, tools, or other specialized devices through variable programmed motions for the performance of a variety of tasks". A NC machining center would be qualified as a robot if one can interpret different types of machining as different functions. Most manufacturing engineers do not consider a NC machining center a robot, even though these machines have a number of similarities. The power drive and controllers of both NC machines and robots can be quite similar. Robots, like NC machine, can be powered by electrical motors, hydraulic systems, or pneumatic systems. Control for either device can be either open-loop or closed-loop. In fact, many of the developments used in robotics have evolved from the NC industry, and many of the manufacturers of robots also manufacture NC machines or NC controllers.

A physical robot is normally composed of a main frame (or arm) with a wrist and some tooling (usually some type of gripper) at the end of the frame. An auxiliary power system may also be included with the robot. A controller with some type of teach pendant, joy-stick, or key-pad is also part of the system. A typical robot system is shown in Figure 4.13.

The history of industrial automation is characterized by periods of rapid change in popular methods. Present use of industrial robots is concentrated in rather simple, repetitive tasks which tend not to require high precision. However, the major growth of the robot population will occur in nonautomotive industries. The most important form of the industrial robot is the mechanical manipulator. Exactly what constitutes an industrial robot is sometimes debated. The distinction lies somewhere in the sophistication of the programmability of the device—if a mechanical device can be programmed to perform a wide variety of applications, it is probably an industrial robot.

Robots are usually characterized by the design of the mechanical system. A robot whose main frame consists of three linear axes is called a Cartesian robot. The Cartesian robot derives its name from the coordinate system. Travel normally takes place linearly in three-space. Some Cartesian robots are constructed like a gantry to minimize deflection along each of the axes. These robots are referred to as gantry robots. Figure 4.14 shows an example of Cartesian robots. These robots behave and can be controlled similarly to conventional three-axis NC machines [1]. Gantry structure are generally the most accurate physical structure for robots. Gantry robots are commonly used for assembly where tight tolerance and exact location are required.

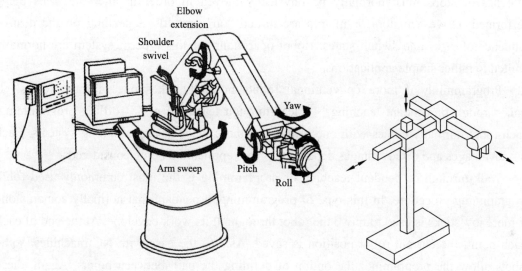

Figure 4.13 A typical robot system   Figure 4.14 A Cartesian robot

A cylindrical robot is composed of two linear axes and one rotary axis. This robot derives its name from the work envelope (the space in which it operates), which is created by moving the axes from limit to limit. Figure 4.15 shows typical cylindrical robots. Cylindrical robots are used for a variety of applications, but most frequently for material-handling operations.

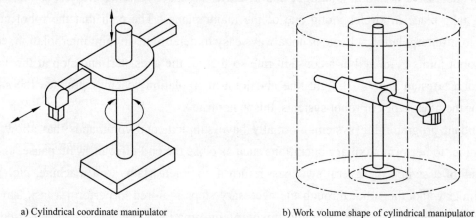

a) Cylindrical coordinate manipulator   b) Work volume shape of cylindrical manipulator

Figure 4.15 Cylindrical robots

4.4.2 Programming a robot

In order for a device to qualify as a robot, it must be easily reprogrammable. Nonprogrammable mechanisms, regardless of their potential flexibility by reassembly or rewiring, do not be qualified as robots [2]. A class of devices that fit this category are fixed or variable-sequence robots. Many of these robots are pneumatically driven. Rather than controlling the robot path, the device is driven to fixed stops or switches via some form of ladder logic. Although the ladder programming qualifies for the definition of a robot, the switches or stops must normally be physically moved in order to alter the tasks being performed. Drive actuators or motors are turned "on" or "off" depending on the desired sequence of tasks and switch states. Robot operations for this type of system are normally limited to rather simple applications.

Programming of more conventional robots normally takes one of three forms: ① walk-through or pendant teaching; ② lead-through teaching; or ③ offline programming. Each robot normally comes with one or more of these types of programming systems. Each has advantages and disadvantages depending on the application being considered.

Walk-through or pendant teaching or programming is the most commonly used robot programming procedure. In this type of programming, a pendant that normally contains one or more joy-sticks is used to move the robot throughout its work envelope. At the end of each teach point, the current robot position is saved. As was the case with NC machines, some robots allow the programmer the option of defining the path between points. Again, these robots are called continuous-path systems. Systems that do not allow the user to specify the path taken are called point-to-point systems. Many continuous-path robots allow the user to define the path to be taken between successive points. That is, the user may define a straight-line, circular, or joint-interpolated path. In a straight-line path, the robots move between successive points in a straight-line in Cartesian space. Circular moves, as the name implies, take place in circles along one of the major planes. The path that the robot takes using a joint-interpolation scheme is not always easy to determine. In joint interpolation, each of the robot joints is moved at a constant rate so that all the axes start and stop at the same time. For Cartesian robots, straight-line and joint-interpolation schemes produce the same path. For the other types of robot systems, this is not true.

Pendant programming systems normally have supplemental commands that allow the programmer to perform auxiliary operations such as close the end-effector, wait, pause, check the status of a switch or several switches, return a complete status to a machine, etc. The programmer walks the robot through the necessary steps required to perform a task, saving each intermediate step along with the auxiliary information. The teach pendant used to program the Fanuc M1 robot is shown in Figure 4.16.

Lead-through programming is one of the simplest programming procedures used to program a robot. As the name implies, the programmer simply physically moves the robot through the required sequence of motions. The robot controller records the position and speed as the programmer leads the robot through the operation. The power is normally shut down while the programmer is leading the robot through the necessary moves, so that the robot will not generate any "glitches" that might injure the operator. Although lead-through programming is the easiest programming method to learn, it does introduce some severe limitations to the robot's application. For instance, when the robot is being led through the operation, the operator carries the weight of the robot. The gears, motor and lead screw may introduce a false resolver reading, so that when the weight of the robot, and perhaps a part, must be supported by the system the actual end-effector position may be significantly different from the position taught to the robot. Another problem with this method is that since the position and speed are recorded as the robot is being led through the desired path, a significant amount of data are generated. This data must be stored and later recalled. Storage and retrieval space and time can cause the programmer problems. Perhaps the major problem associated with lead-through programming is that the human who leads the robot through the process is capable of only finite accuracy and may introduce inconsistencies into the process. Human-induced errors and inaccuracies eliminate some of the major advantages of using robots.

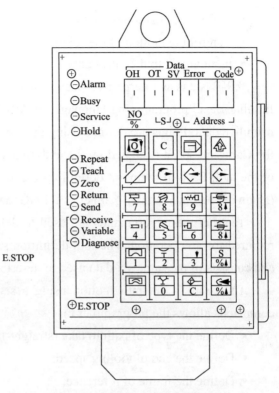

Figure 4.16　Fanuc M1 teach pendant

Offline programming for robots is a relatively new technology that provides several advantages over both lead-through and teach programming. The principles of offline programming are similar to using an offline language such as APT for NC programming. Several languages have been developed at major universities as well in industry throughout the US. Examples of these languages include VAL created by Unimation, AR-Basic by American Robot Corp., ARM-Basic by Microbot, Inc. and AMI by IBM. To illustrate offline programming, AR-Basic will be used. AR-Basic allows the user to:

- Define the position of the robot;
- Control the motion of the robot;
- Input and output control data.

AR-Basic is an interpretive Basis system that employs many of the same functions as the familiar Basic programming language. In AR-Basic, points and tools are defined as a set of primitive data. Points are defined using the convention: $X, Y, Z, R, P, Y,$ where $X, Y,$ and Z are the Cartesian space occupied by the end-effector, and $R, P,$ and Y are the roll, pitch and yaw of the tool. Each of the point definitions can be specified as either absolute or relative points (again defined in a similar manner as for NC machines).

Tool-definition commands are used to define the location of any tooling that might be required for an operation. The tool definition specifies the midpoint of the robot's faceplate, and consists of the same six data used to describe a point.

The robot is set into motion using a set of motion control commands. The motion commands allows the programmer to:

- Define the type of path to take (straight-line, circular, or joint-coordinated);
- Define the end of tooling speed;
- Define the frame of reference;
- Describe the current tool tip.

AR-Basic also allows the programmer to input and output data to any device to which the robot is interfaced. Analog and digital data can be sent to A/D converters, D/A converters, parallel, or serial I/O ports. Table 4.3 contains point and tool definition examples. Table 4.4 contains AR-Basic examples for motion control.

Table 4.3 Point and tool definition statements

TEACH_ABS POINT_1	!Define an Absolute Point based on the position of the robot's tool tip
POINT_2=DEF_REL(30,-5,4,,45,180)	!Define a Relative Point by specifying the six components
REL_POINT_1=CONVERT_REL(POINT_1)	!Convert an Absolute Point definition to a Relative Point definition (based on the current frame)
DRILL_BIT=DEF_TOOL(8)	!Define a simple 8-in. straight Tool (all unspecified parameters default to 0)
PRINT POINT_2.PITCH	Print on the screen the Pitch component of POINT_2
PINT_1.Y =SQR(ABS(-144))	Set the Y component of POINT_1 to 12

Table 4.4 Motion-control command statements

SET_SPEED TO 30	Set Tool tip speed to 30 in/s
SET_MOTION TO STRAIGHT	Specify straight line motion
SET_TOOL TO DRILL_BIT	Tool defined in last example
SET_FRAME TO POINT_1	Set current frame using an Absolute Point definition

(continued table)

MOVE TO REL_POINT_1	Move to position specified by Relative Point definition (relative to frame specified by Absolute Point POINT_1
MOVE TO POINT_2, AT SPEED 20, USING JOINT MOTION	Move to POINT_2, temporarily overriding the current default speed and motion style
MOVE $XYZ_TABLE TO POINT_1	Move an independently defined *XYZ* table

Words and Expressions

circular move [ˈsɜːkjʊlə muːv]	圆弧运动，在某一主平面上沿圆弧运动
joint interpolation [dʒɔɪnt ɪnˌtəpəʊˈleɪʃən]	结点插补

Notes

[1] 本句可译为：这些机器人的动作的控制都类似于传统的三坐标数控机床。

[2] 本句可译为：不可编程的机构，无论其通过重新装配或再接线可实现的潜在柔性有多大，也不能算作是机器人。

Questions for Discussion

How to think about the statement: "a robot must be easily reprogrammable"?

Chapter 5 Mechatronic Engineering

5.1 Advanced Control and Automation of Mechatronic Systems

5.1.1 Control Systems

Control systems are an integral part of modern society. Numerous applications are all around us: The rockets fire, and the space shuttle lifts off to earth orbit; in splashing cooling water, a metallic part is automatically machined; a self-guided vehicle delivering material to workstations in an aerospace assembly plant glides along the floor seeking its destination, etc.

(1) Control system definition

A control system consists of subsystems and processes (or plants) assembled for the purpose of obtaining a desired output with desired performance, given a specified input. Figure 5.1 shows a control system in its simplest form, where the input represents a desired output. For example, consider an elevator. When the fourth-floor button is pressed on the first floor, the elevator rises to the fourth floor with a speed and floor-leveling accuracy designed for passenger comfort. The push of the fourth-floor button is an input that represents our desired output, shown as a step function in Figure 5.2. The performance of the elevator can be seen from the elevator response curve in the figure. Two major measures of performance are apparent: ① the transient response and ② the steady-state error. In this example, passenger comfort and passenger patience are dependent upon the transient response. If this response is too fast, passenger comfort is sacrificed; if too slow, passenger patience is sacrificed. The steady-state error is another important performance specification since passenger safety and convenience would be sacrificed if the elevator did not properly level.

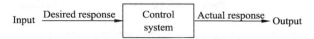

Figure 5.1 Simplified description of a control system

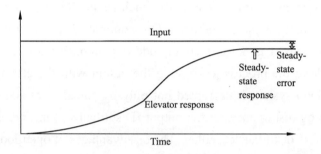

Figure 5.2 Elevator response

(2) The classes of control systems

There are two common classes of control action: open loop and closed loop.

1) Open-loop systems

In an open-loop control system, the control action from the controller is independent of the process variable. Open-loop systems are simply commanded by the input. For example, toasters are open-loop systems, as anyone with burnt toast can attest. The controlled variable (output) of a toaster is the color of the toast. The device is designed with the assumption that the toast will be darker the longer it is subjected to heat. The toaster does not measure the color of the toast; it does not correct for the fact that the toast is rye, white, or sourdough, nor does it correct for the fact that toast comes in different thicknesses. Other examples of open-loop systems are mechanical systems consisting of a mass, spring, and damper with a constant force positioning the mass. The greater the force, the greater the displacement.

Open–loop systems

A general open-loop system is shown in Figure 5.3. It starts with a subsystem called an input transducer, which converts the form of the input to that used by the controller. The controller drives a process or a plant. The input is sometimes called the reference, while the output can be called the controlled variable. Other signals, such as disturbances, are shown added to the controller and process outputs via summing junctions, which yield the algebraic sum of their input signals using associated signs. For example, the plant can be a furnace or air conditioning system, where the output variable is temperature. The controller in a heating system consists of fuel valves and the electrical system that operates the valves.

Figure 5.3   A block diagram of a open-loop control system

The distinguishing characteristic of an open-loop system is that it cannot compensate for any disturbances that add to the controller's driving signal (Disturbance 1 in Figure 5.3). The system position will change with a disturbance, such as an additional force, and the system will not detect or correct for the disturbance. For example, if the controller is an electronic amplifier and Disturbance 1 is noise, then any additive amplifier noise at the first summing junction will also drive the process, corrupting the output with the effect of the noise. The output of an open-loop system is corrupted not only by signals that add to the controller's commands but also by disturbances at the output (Disturbance 2 in Figure 5.3). The system cannot correct for these disturbances, either. The disadvantages of open-loop systems, namely sensitivity to disturbances and inability to correct for these disturbances, may be overcome in closed-loop systems.

2) Closed-loop systems

Closed–loop systems

In a closed-loop system, the control action from the controller is dependent on the desired and actual process variable. In the case of the boiler analogy, this would utilize a thermostat to monitor the building temperature, and feedback a signal to ensure the controller output maintains the building temperature to that set on the thermostat. A closed loop controller has a feedback loop which ensures the controller exerts a control action to control a process variable at the same value as the reference signal. For this reason, closed-loop controllers are also called feedback controllers.

The general architecture of a closed-loop system is shown in Figure 5.4. The input transducer converts the form of the input to the form used by the controller. An output transducer, or sensor, measures the output response and converts it into the form used by the controller. For example, if the controller uses electrical signals to operate the valves of a temperature control system, the input position and the output temperature are converted to electrical signals. The input position can be converted to a voltage by a potentiometer, a variable resistor, and the output temperature can be converted to a voltage by a thermistor, a device whose electrical resistance changes with temperature. The first summing junction algebraically adds the signal from the input to the signal from the output, which arrives via the feedback path, the return path from the output to the summing junction. In Figure 5.4, the output signal is subtracted from the input signal. The result is generally called the actuating signal. However, in systems where both the input and output transducers have unity gain (that is, the transducer amplifies its input by Disturbance 1 in Figure 5.4), the actuating signal's value is equal to the actual difference between the input and the output. Under this condition, the actuating signal is called the error.

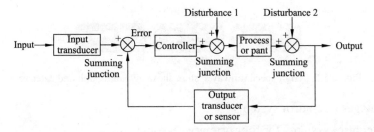

Figure 5.4   A block diagram of a closed-loop control system

Closed-loop systems, then, have the obvious advantage of greater accuracy than open-loop systems. They are less sensitive to noise, disturbances, and changes in the environment. Transient response and steady-state error can be controlled more conveniently and with greater flexibility in closed-loop systems, often by a simple adjustment of gain (amplification) in the loop and sometimes by redesigning the controller. We refer to the redesign as compensating the system and to the resulting hardware as a compensator. The closed-loop system compensates for disturbances by measuring the output response, feeding that measurement back through a feedback path, and comparing that response to the input at the summing junction. If there is any difference between the two responses, the system drives the plant, via the actuating signal, to make a correction. If there is no difference, the system does not drive the plant, since the plant's response is already the desired response.

On the other hand, closed-loop systems are more complex and expensive than open-loop systems. A standard, open-loop toaster serves as an example: It is simple and inexpensive. A closed-loop toaster oven is more complex and more expensive since it has to measure both color (through light reflectivity) and humidity inside the toaster oven. Thus, the control systems engineer must consider the trade-off between the simplicity and low cost of an open-loop system and the accuracy and higher cost of a closed-loop system.

Some high speed railway systems are powered by electricity supplied to a pantograph on the train's roof from a catenary overhead, as shown in Figure 5.5. The force applied by the pantograph to the catenary is regulated to avoid loss of contact due to excessive transient motion. A proposed method to regulate the force uses a closed-loop feedback system, whereby a force, F_{up}, is applied to the bottom of the pantograph, resulting in an output force applied to the catenary at the top. The contact between the head of the pantograph and the catenary is represented by a spring. The output force is proportional to the displacement of this spring, which is the difference between the catenary and pantograph head vertical positions.

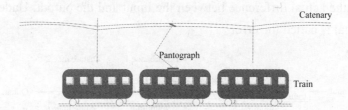

Figure 5.5 High speed railway system showing pantograph and catenary

(3) Advantages of control systems

We build control systems for four primary reasons:
- Power amplification;
- Remote control;
- Convenience of input form;
- Compensation for disturbances.

With control systems we can move large equipment with precision that would otherwise be impossible. We can point huge antennas toward the farthest reaches of the universe to pick up faint radio signals; controlling these antennas by hand would be impossible. Because of control systems, elevators carry us quickly to our destination, automatically stopping at the right floor (shown in Figure 5.2). We alone could not provide the power required for the load and the speed; motors provide the power, and control systems regulate the position and speed.

A control system can produce the needed power amplification, or power gain. Robots designed by control system principles can compensate for human disabilities. Control systems are also useful in remote or dangerous locations. For example, a remote-controlled robot arm can be used to pick up material in a radioactive environment. Figure 5.6 shows a robot arm designed to work in contaminated environments. Control systems can also be used to provide convenience by changing the form of the input. For example, in a temperature control system, the input is a position on a thermostat. The output is heat. Thus, a convenient position input yields a desired thermal output. Another advantage of a control system is the ability to compensate for disturbances. Typically, we control such variables as temperature in thermal systems, position and velocity in mechanical systems, and voltage, current, or frequency in electrical systems. The system must be able to yield the correct output even with a disturbance.

Figure 5.6 The remote-controlled robot arm

Control systems contribute to every aspect of modern society. In our homes we find them in everything from toasters to heating systems to VCRs. Control systems also have widespread applications in science and industry, from steering ships and planes to guiding missiles and the space shuttle. Control systems also exist naturally; our bodies contain numerous control systems. Control systems are used where power gain, remote control, or conversion of the form of the input is required.

A control system has an input, a pant, and an output. Control systems can be open loop or closed loop. Open-loop systems do not monitor or correct the output for disturbances; however, they are simpler and less expensive than closed-loop systems. Closed-loop systems monitor the output and compare it to the input. If an error is detected, the system corrects the output and hence corrects the effects of disturbances.

Control systems analysis and design focuses on three primary objectives:
- Achieving stability;
- Producing the desired transient response;
- Reducing steady-state errors.

A system must be stable in order to produce the proper transient and steady state response. Transient response is important because it affects the speed of the system and influences human patience and comfort, not to mention mechanical stress. Steady-state response determines the accuracy of the control system; it governs how closely the output matches the desired response.

5.1.2 PID Controller

PID (proportional integral derivative) controller is a particular control structure that has become almost universally used in industrial control. PID is a "closed loop" system, meaning that the results of the output drive system are known to the input control system. Closed loop systems are vitally important to process control because almost no control solution operates in a perfectly linear and predictable environment. A closed loop system simplifies process control by changing the nature of the problem. Rather than indirectly factoring in all the possible variables and quantify their individual and cumulative effects on the output, you simply control the output. All the other things take care of themselves.

PID controller

It has been around for decades, and while there have been attempts to improve it by adding more control information, or replace it all together, it is still the bedrock from which all process control springs. Yet, as common and popular as it is, its functioning seems to be taken more as a "black box" by hobbyists who may basically understand it but who are a bit intimidated to delve too deeply into its operation. While it's behavior is founded in math that may be a bit complex for some beginners, it can be explained in conceptual terms that are easy to understand.

(1) PID structure

Consider the simple SISO control loop shown in Figure 5.7.

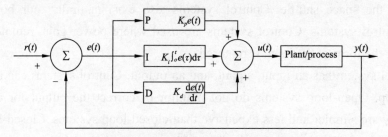

Figure 5.7 A block diagram of a PID controller in a feedback loop

The distinguishing feature of the PID controller is the ability to use the three control terms of proportional, integral and derivative influence on the controller output to apply accurate and optimal control. The block diagram on the right shows the principles of how these terms are generated and applied. It shows a PID controller, which continuously calculates an error value $e(t)=r(t)-y(t)$ as the difference between a desired value $r(t)$ and a measured process variable $y(t)$, and applies a correction based on proportional, integral, and derivative terms. The controller attempts to minimize the error over time by adjustment of a control variable $u(t)$, such as the opening of a control valve, to a new value determined by a weighted sum of the control terms.

The overall control function can be expressed mathematically as follows:

$$u(t) = K_p e(t) + K_i \int_0^t e(t) dt + K_d \frac{de(t)}{dt} \tag{5.1}$$

where K_p, K_i and K_d, all non-negative, denote the coefficients for the proportional, integral and derivative terms respectively (sometimes denoted P, I and D)

The traditional expressions for PI and PID controllers can be designed by their transfer functions, relating error $E(s)=Y(s)-R(s)$ and controller output $U(s)$ as follows:

$$C_P(s) = K_p \tag{5.2}$$

$$C_{PI}(s) = K_p \left(1 + \frac{1}{T_r s}\right) \tag{5.3}$$

$$C_{PD}(s) = K_p \left(1 + \frac{T_d s}{\tau_d s + 1}\right) \tag{5.4}$$

$$C_{PID}(s) = K_p \left(1 + \frac{1}{T_r s} + \frac{T_d s}{\tau_d s + 1}\right) \tag{5.5}$$

where T_r and T_d are known as the reset time and derivative time respectively.

As seen from (5.2) to (5.5), the members of this family include, in different combinations, three control modes or actions proportional (P), integral (I), and derivation (D).

Caution must be exercised when applying PID tuning rules, as there are a member of other parameterizations. Equation 5.5 is known as the standard form. An alternative, series form, is as follows:

$$C_{series}(s) = K_s \left(1 + \frac{I_s}{s}\right)\left(1 + \frac{D_s s}{\gamma_s D_s s + 1}\right) \tag{5.6}$$

The parallel form is as follows:

$$C_{parallel}(s) = K_p + \frac{I_p}{s} + \frac{D_p s}{\gamma_p D_p + 1} \tag{5.7}$$

Terminology, such as P-gain are not uniquely defined and can refer to either K_s in Equation 5.6, K_p in (5.7) or K_p in (5.5). It is therefore important to know which of these parameterizations any one particular technique refers to and, if implemented in a different form, to transform the parameters appropriately. Before the PID was recognized as simply a second-order controller, PID tuning was viewed in terms of the P, I and D parameters.

The balance of these effects is achieved by "loop tuning" to produce the optimal control function. The tuning constants are shown below as "K" and must be derived for each control application, as they depend on the response characteristics of the complete loop external to the controller. These are dependent on the behavior of the measuring sensor, the final control element (such as a control valve), any control signal delays and the process itself. Approximate values of constants can usually be initially entered knowing the type of application, but they are normally refined, or tuned, by "bumping" the process in practice by introducing a desired value change and observing the system response.

The mathematical model and practical loop above both use a "direct" control action for all the terms, which means an increasing positive error results in an increasing positive control output for the summed terms to apply correction. However, the output is called "reverse" acting if it is necessary to apply negative corrective action. For instance, if the valve in the flow loop was 100%～0 valve opening for 0～100% control output-meaning that the controller action has to be reversed. Some process control schemes and final control elements require this reverse action. An example would be a valve for cooling water, where the fail-safe mode, in the case of loss of signal, would be 100% opening of the valve; therefore 0 controller output needs to cause 100% valve opening.

Although a PID controller has three control terms, some applications use only one or two terms to provide the appropriate control. This is achieved by setting the unused parameters to zero and is called a PI, PD, P or I controller in the absence of the other control actions. PI controllers are fairly common, since derivative action is sensitive to measurement noise, whereas the absence of an integral term may prevent the system from reaching its target value. They have proven to be robust in the control of many important applications. The simplicity of these controllers is also there weakness: it limits the range of plants that they can control satisfactorily. Indeed, there exists a set of unstable plants which cannot be stabilized with any member of the PID family.

(2) Proportional control

Term "P" is proportional to the current value of the error $e(t)$. For example, if the error is large and positive, the control output will be proportionately large and positive, taking into account the gain factor "K". Using proportional control alone in a process with compensation such as temperature control, will result in an error between the desired value and the actual process value, because it requires an error to generate the proportional response. If there is no error, there is no corrective response.

Proportional action provide a contribution which depends on the instantaneous value of the control error. A proportional controller can control any stable plant, but it provides limited performance and nonzero steady-state errors. This latter limitation is due to the fact that its frequency response is bounded for all frequencies.

It has also been traditional to use the expression proportional band (PB) to describe the proportional action. The equivalence is

$$PB[\%] = \frac{100[\%]}{K_p} \tag{5.8}$$

The proportional band is defined as the error required (as a percentage of full scale) to yield a 100% change in the controller output.

Unlike the oven, controlling more demanding devices, like a motor or motorized vehicle, by simply turning it on and turning it off is probably a bad idea. For instance, a typical oven set at 350 degrees will drop to about 325 degrees before it turns on, then shoot up to about 375 before it turns off. This is a huge range of error. Imagine the cruise control system of an automobile set for 60 miles an hour speeding up to 75 miles an hour and dropping down to 45 before it speeds up to 75 again.

Proportional control works mostly on the notion of "error". The error is how far or behind the device is from where it should be. A greater error requires a greater amount of force to overcome it. A smaller error requires a smaller amount of force. Theoretically, a proportional system in operation is always in an "error" position. If the error were zero, the output would also be zero.

The problem with a strictly proportional control of devices is that the error in speed is only a single measurement, a snapshot in time. A simple proportional system does not consider if the system is gaining or losing on its target or the rate at which it is gaining or losing. This is the "differential" and "integral" information in "PID", and if ignored will cause the control system to respond poorly. Over compensating when it is catching up, and under compensating when it is losing ground.

(3) Integral control

Term "I" accounts for past values of the error $e(t)$ and integrates them over time to produce the "I" term. For example, if there is a residual error after the application of proportional control, the integral term seeks to eliminate the residual error by adding a control effect due to the historic cumulative value of the error. When the error is eliminated, the integral term will cease to grow. This will result in the proportional effect diminishing as the error decreases, but this is compensated for by the growing integral effect.

The integral, or the "I" in PID, is an accumulation of error, it compensates for previous behavior, it is sort of the ghost of error past. Each time the error is calculated, it is added to a number which represents the total accumulated error for the control process. As this value increases or decreases from each error calculation, it is added to the proportional power output. Basically, even if a motor's current error is zero, (it's going the correct speed) the integrated error can still affect the motors power to make up for previous errors in overall motion, an important step if the system is working in conjunction with other systems, for instance the classic two wheel drive robot where if the two opposing wheels move different distances it turns.

Integral action on the other hand, gives a controller output that is proportional to the accumulated error, which implies that it is a slow reaction control mode. This characteristic is also evident in its low-pass frequency response. The integral mode plays a fundamental role in achieving perfect plant inversion at $\omega=0$. This force the steady-state error to zero in the presence of a step reference and disturbance. The integral mode, viewed in isolation, has two major shortcomings: its pole at the origin is detrimental to loop stability and it also gives rise to the undesirable effect (in the presence of actuator saturation) known as wind-up.

(4) Differential control

Term "D" is the rate at which the error is changing, it adjusts for anticipated behavior. If the current error is less than the previous error, the system is gaining on where it should be. If the current error is greater than the previous error, then the system is losing. The differential information can be used to modify the proportional control even further to prevent over-shoot or make it catch up more quickly. The differential value is added into the proportional output along with the integral value to augment the effects of the proportional or integral values.

Differential control is sometimes called "anticipatory control," as it is effectively seeking to reduce the effect of the error by exerting a control influence generated by the rate of error change. Derivative action acts on the rate of change of the control error. Consequently, it is a fast mode which ultimately disappears in the presence of constant errors. The more rapid the change, the greater the controlling or dampening effect.

It is sometimes referred to as a predictive mode, because of its dependence on the error trend. The main limitation of the derivative mode, viewed in isolation, is its tendency to yield large control signals in response to high frequency control errors, such as errors induced by set-point changes or measurement noise.

Its implementation requires properness of the transfer functions, so a pole typically added to the derivative, as is evident in Equation 5.4 and Equation 5.5. In the absence of other constraints, the additional time constant τ_D is normally chosen such that $0.1T_d \leqslant \tau_D \leqslant 0.2T_d$. This constant is called the derivative time constant; the smaller it is, the larger the frequency range over which the filtered derivative approximates the exact derivative, with equality in the limit.

$$\lim_{\tau_D \to 0} u_{\text{PID}}(t) = K_p e(t) + \frac{K_p}{T_r} \int_{t_0}^{t} e(\tau) \mathrm{d}\tau + K_p T_d \frac{\mathrm{d}e(t)}{\mathrm{d}t} \qquad (5.9)$$

The classical argument to choose $\tau_D \neq 0$ was, apart from ensuring that the controller be proper, to attenuate high-frequency noise. The latter point is illustrated in Figure 5.8, which shows that the filtered derivative approximates the exact derivative well at frequencies up to $\frac{1}{\tau_D}(\text{rad/s})$, but that it has finite gain at high frequencies, whereas an exact derivative infinite gain.

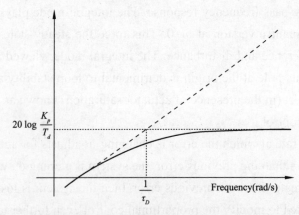

Figure 5.8 Bode magnitude plot of exact derivative (dashed) and filtered derivative (solid)

Because $\tau_D \neq 0$ was largely seen as a necessary evil, i.e., as a necessary departure from a pure proportional, integral, and derivative action, almost all industrial PID controllers once set τ_D as a fixed fraction of T_D, rather than viewing it as an independent design parameter in its own right, Since then it has become clear, however, that the time constant of the derivative is an important degree of freedom available to the designer.

(5) Application of PID controller

In all fields of engineering, many different solutions can be presented for the same problem. The goal of an engineer is twofold. The first and most distinct goal is to identify and design possible solutions to a complex problem. The second goal, which is perhaps less obvious, is to choose which solution is best suitable to most economically meet project specifications. This project deals with the general field of electrical engineering. Its primary goal is to implement a digital "proportional integral derivative" controller in an existing DC motor system with this unwritten economical specification that cost must be minimized.

An existing DC motor system will be used as our closed loop system "plant". This system utilized an 80C515 microcontroller to control the speed of a 30-Volt DC motor by pulse width modulation. This system was considered to be an "open-loop" system because no information was fed back to the microprocessor with the express purpose of improving system performance. Two major problems arose due to the open-loop nature of this system. The first problem was system non-linearity that arose due to the nature of the hardware. The second problem was system unreliability when the DC motor load was varied. This project will attempt to address both of these problems by means of a PID controller.

One drawback of PID control is overall complexity. This results in very expensive means of implementing a digital version of a PID controller. Of the many possibilities, digital signal processors (DSP's) are the most widely used to solve this problem, however other possibilities exist which may be more cost-effective. Implementation of any complex digital controller must be done by means of some form of computer. Typical microcontrollers, while cheap, do not normally provide enough processing power to effectively perform all but the most simple calculations real-time. Digital signal processors, on the other hand, are designed to implement complex algorithms quickly. The major drawback of DSP's, however, is cost. This project will attempt to find a median between these two extremes of performance and cost. The proposed solution is to design a special-purpose computer whose only purpose is to quickly execute the complex PID algorithm. This computer will be designed using the IEEE 1076-1987 standard known as VHDL (Very High Speed Integrated Circuit Hardware Description Language)[1], and will be implemented on an FPGA (Field Programmable Gate Array)[2].

The input of the system is speed command signal and the outputs are motor shaft velocity and system display. The speed command signal is inputted to the microcontroller, which in turn controls the speed of the motor. The motor shaft velocity is simply the speed at which the motor turns. The system display provides vital information about the system including both the current PWM command signal setting and the current motor speed measured in revolutions per minute (RPM).

The top-level system block diagram (shown in Figure 5.9) illustrates the digital PID controller and DC motor system in a closed loop configuration. The motor shaft velocity will be fed back and confirmed with the speed command signal to drive the system by means of an error signal. This more detailed look at the top-level block diagram of the overall system (shown in Figure 5.10) illustrates the basic forms that the block transfer functions will follow. As more detailed models of the PID Controller and DC Motor are derived, this diagram will increase in complexity. Also, a method of creating an error signal based off the speed command signal and motor shaft velocity signal will need to be developed.

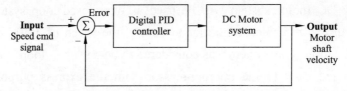

Figure 5.9   Top-level system block diagram

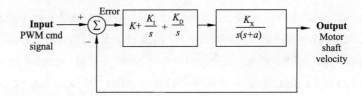

Figure 5.10   Detailed top-level block diagram

While PID controllers are applicable to many control problems, and often perform satisfactorily without any improvements or only coarse tuning, they can perform poorly in some applications, and do not in general provide optimal control. The fundamental difficulty with PID control is that it is a feedback control system, with constant parameters, and no direct knowledge of the process, and thus overall performance is reactive and a compromise. While PID control is the best controller in an observer without a model of the process, better performance can be obtained by overtly modeling the actor of the process without resorting to an observer. PID controllers, when used alone, can give poor performance when the PID loop gains must be reduced so that the control system does not overshoot, oscillate or hunt about the control setpoint value. They also have difficulties in the presence of nonlinearities, may trade-off regulation versus response time, do not react to changing process behavior (say, the process changes after it has warmed up), and have lag in responding to large disturbances. The most significant improvement is to incorporate feed-forward control with knowledge about the system, and using the PID only to control error. Alternatively, PIDs can be modified in more minor ways, such as by changing the parameters (either gain scheduling in different use cases or adaptively modifying them based on performance), improving measurement (higher sampling rate, precision, and accuracy, and low-pass filtering if necessary), or cascading multiple PID controllers.

The use of the PID algorithm does not guarantee optimal control of the system or its control stability. Situations may occur where there are excessive delays: the measurement of the process value is delayed, or the control action does not apply quickly enough. In these cases lead-lag compensation is required to be effective. The response of the controller can be described in terms of its responsiveness to an error, the degree to which the system overshoots a setpoint, and the degree of any system oscillation. But the PID controller is broadly applicable, since it relies only on the response of the measured process variable, not on knowledge or a model of the underlying process.

5.1.3 Brief Introduction of Advanced Control Methods

Advanced control methods are rapidly becoming an established field of education and research, not simply because of its pedagogical importance but also in view of its tremendous success in industrial applications. Fuzzy logic, neural networks, and evolutionary computing have provided important tools and techniques for system control. Specifically, the field of advanced control is fertile with techniques of computational intelligence; particularly, soft computing, that are, by and large, based on a fuzzy-neural-evolutionary framework. Advanced control seeks to achieve good performance in machines, industrial processes, consumer products, and other systems, by using control approaches that, in a loose sense, tend to mimic direct control by experienced humans. Many of these techniques can learn, adapt to compensate for parameter changes and disturbances, and are able to provide satisfactory control even in incompletely-known and unfamiliar situations.

The advanced control methods sometimes refer to intelligent control methods. The term "intelligent control" may be loosely used to denote a control technique that can be carried out using the "intelligence" of a human who is knowledgeable in the particular domain of control. In this definition, constraints pertaining to limitations of sensory and actuation capabilities and information processing speeds of humans are not considered. It follows that if a human in the control loop can properly control a plant, then that system would be a good candidate for intelligent control. Information abstraction and knowledge-based decision making that incorporates abstracted information are considered important in intelligent control.

Unlike conventional control, intelligent control techniques possess capabilities of effectively dealing with incomplete information concerning the plant and its environment, and unexpected or unfamiliar conditions. The term "adaptive control" is used to denote a class of control techniques where the parameters of the controller are changed (adapted) during control, utilizing observations on the plant (i.e. with sensory feedback), to compensate for parameter changes, other disturbances, and unknown factors of the plant. Combining these two terms, one may view "intelligent adaptive control" as those techniques that rely on intelligent control for proper operation of a plant, particularly in the presence of parameter changes and unknown disturbances.

There are several classic advanced control methods, such as adaptive control, fuzzy logic control, neural network control, etc.

(1) Adaptive control

Adaptive control

Adaptive control is the control method used by a controller which must adapt to a controlled system with parameters which vary, or are initially uncertain. For example, as an aircraft flies, its mass will slowly decrease as a result of fuel consumption; a control law is needed that adapts itself to such changing conditions. Adaptive control does not need a priori information about the bounds on these uncertain or time-varying parameters and is concerned with control law changing itself. The basic structure of an adaptive controller is shown in Figure 5.11.

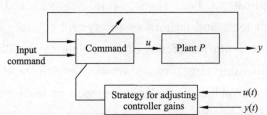

Figure 5.11　Adaptive controller structure with adjustable controller gains

Adaptive control systems are time varying and nonlinear, thus more challenging to analyze and understand than traditional linear time invariant controllers. The stability proofs are often long and technical and possibly distracting to readers who prefer to focus on the design and implementation of adaptive control. Adaptive control is an active field in the design of control systems to deal with uncertainties. The key difference between adaptive controllers and linear controllers is the adaptive controller's ability to adjust itself to handle unknown model uncertainties. Adaptive control is roughly divided into two categories: direct and indirect. Indirect methods estimate the parameters in the plant and further use the estimated model information to adjust the controller. Direct methods are ones wherein the estimated parameters are those directly used in the adaptive controller.

Much effort has been placed in adaptive control in both theory and applications. Theory-wise, new controller design techniques are introduced to handle nonlinear and time-varying uncertainties. Broader systems with larger nonlinear uncertainties can be covered by these developments. As a result, adaptive control finds use in various real world applications.

There have been many experiments on adaptive control in laboratories and industry. The rapid progress in microelectronics was a strong stimulation. Interaction between theory and experimentation resulted in a vigorous development of the field. As a result, adaptive controllers started to appear commercially in the early 1980s. This development is now accelerating. One result is that virtually all single-loop controllers that are commercially available today allow adaptive techniques of some form. The primary reason for introducing adaptive control was to obtain controllers that could adapt to changes in process dynamics and disturbance characteristics. It has been found that adaptive techniques can also be used to provide automatic tuning of controllers.

1) Model-reference adaptive systems

The model-reference adaptive system (MRAS) was originally proposed to solve a problem in which the performance specifications are given in terms of a reference model. This model tells how the process output ideally should respond to the command signal. A block diagram of the system is shown in Figure 5.12. The controller can be thought of as consisting of two loops. The inner loop is an ordinary feedback loop composed of the process and the controller. The outer loop adjusts the controller parameters in such a way that the error, which is the difference between process output y and model output y_m, is small. The MRAS was originally introduced for flight control. In this case the reference model describes the desired response of the aircraft to joystick motions.

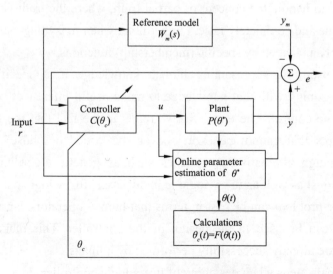

Figure 5.12 Indirect MRAC

The key problem with MRAS is to determine the adjustment mechanism so that a stable system, which brings the error to zero, is obtained. This problem is nontrivial. The following parameter adjustment mechanism was used in the original MRAS:

$$\frac{d\theta}{dt} = -\gamma e \frac{\partial e}{\partial \theta} \quad (5.10)$$

In this equation, $e = y - y_m$ denotes the model error and θ is a controller parameter. The quantity $\frac{\partial e}{\partial \theta}$ is the sensitivity derivative of the error with respect to parameter θ. The parameter γ determines the adaptation rate.

In practice it is necessary to make approximations to obtain the sensitivity derivative. The rule can be regarded as a gradient scheme to minimize the squared error e^2.

2) Application of adaptive controller in flight control

The dynamics of an airplane change significantly with speed, altitude, angle of attack, and so on. Control systems such as autopilots and stability augmentation systems were used early. These systems were based on linear feedback with constant coefficients. This worked well when speeds and altitudes were low, but difficulties were encountered with increasing speed and altitude. The problems became very pronounced at supersonic flight. Flight control was one of the strong driving forces for the early development of adaptive control.

(2) Fuzzy logic control

Fuzzy logic control

The term fuzzy logic was introduced with the 1965 proposal of fuzzy set theory by Zadeh. Fuzzy logic has been applied to many fields, from control theory to artificial intelligence. By contrast, in Boolean logic, the truth values of variables may only be the integer values 0 or 1. Fuzzy logic employed to handle the concept of partial truth, where the truth value may range between completely true and completely false. Furthermore, when fuzzy linguistic variables are used, the degrees may be managed by specific (membership) functions.

Fuzzy logic is useful in representing human knowledge in a specific domain of application, and in reasoning with that knowledge to make useful inferences or actions. It is widely used in machine control. The term "fuzzy" refers to the fact that the logic involved can deal with concepts that cannot be expressed as the "true" or "false" but rather as "partially true". Although alternative approaches such as genetic algorithms and neural networks can perform just as well as fuzzy logic in many cases, fuzzy logic has the advantage that the solution to the problem can be cast in terms that human operators can understand, so that their experience can be used in the design of the controller. This makes it easier to mechanize tasks that are already successfully performed by humans.

There are three main steps when designing of fuzzy logic controller.

1) Fuzzify a variable

To fuzzify the input variable, we often use three kinds of fuzzy sets as follows:

e={NB,NS,ZO,PS,PB}
e={NB,NM,NS,ZO,PS,PM,PB}
e={NB,NM,NS,NZ,PZ,PS,PM,PB}

For example, a fuzzy system with eight fuzzy sets using triangle MF is shown in Figure 5.13.

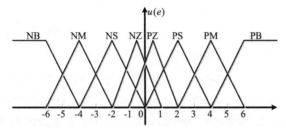

Figure 5.13　Eight triangle fuzzy sets

2) Rule base

Rule base consists of several fuzzy rules. For example, in fuzzy logic control, we can design rule base with two fuzzy rules as follows:

R1 : IF E is NB and EC is NB then U is PB
R2 : IF E is NB and EC is NS then U is PM

where E represents error, EC represents error change, U represents control input.

3) Fuzzy inference and defuzzy

From the rule base, we can get fuzzy relation matrix R. If we have new premise information A and B, we can get a new result:

$$C = (A \times B) \cdot R \qquad (5.11)$$

The conclusion matrix C is fuzzy vector, and it must be defuzzified to exact value for practical use[3].

4) Fuzzy logic control for washing machine

For the washing machine, how to set washing time is an important question. Fuzzy logic control is an important method for the washing time setting of the washing machine, which can be described as several steps as follows:

Consider washing time control of the washing machine, and we can design a fuzzy controller with two inputs and one output. According to our experience, we choose mud and axunge as the inputs and choose washing time as the output.

According to experience, we can define three fuzzy sets for mud and axunge, respectively, and define five fuzzy sets for washing time.

- Consider MF design of mud, we can define three fuzzy sets: SD (mud small), MD (mud middle), and LD (mud large), the range of mud is in [0, 100].
- Consider axunge, we can define three fuzzy sets: NG (no axunge), MG (middle axunge), and LG (large axunge), the range value of axunge is set as [0, 100].

- For washing time, five fuzzy sets are used: VS (very small), S (small), M (middle), L (long), and VL (very long), and the value is in the range of [0, 60].

According to experience, we can design fuzzy rules. The input is mud and axunge, and the output is washing time. If we design three membership functions for each input, then we can design nine rules.

- The format of the rule is "IF Mud is A AND Axunge is B THEN Washing time is C".

(3) Artificial neural network control

Artificial neural network control

Artificial neural networks represent a massively connected network of computational "neurons." By adjusting a set of weighting parameters of a neural network (NN), it may be "trained" to approximate an arbitrary nonlinear function to a required degree of accuracy. Biological analogy here is the neuronal architecture of a human brain.

The successful operation of an autonomous machine depends on its ability to cope with a variety of unexpected and possibly unfamiliar events in its operating environment, perhaps relying on incomplete information. Such an autonomous machine would only need to be presented a goal; it would achieve its objective through continuous interaction with its environment and automatic feedback about its response. In fact, this is an essential part of "learning". By enabling machines to possess such a level of autonomy, they would be able to learn higher-level cognitive tasks that are not easily handled by existing machines. In designing a machine to emulate the capabilities found in biological controls, which depend on "intelligence". Some experience on the structural, functional, and behavioral biological neural systems would be valuable. Neural networks have a great potential in the realm of nonlinear control problems. A significant characteristic of neural networks is their ability to approximate arbitrary nonlinear functions. This ability of neural networks has made them useful in modeling nonlinear systems.

A neural network controller, in general, performs a specific task for adaptive control, with the controller taking the form of a multilayer network, and adaptable parameters being defined as the adjustable weights. In general, neural networks represent parallel-distributed processing structures, which make them prime candidates for use in multi-variable control systems. The neural-network approach defines the problem of control as the mapping of measured signals of system "change" into calculated "control actions", as shown in Figure 5.14.

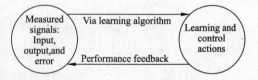

Figure 5.14 Representation of learning and control actions in a neural network approach

The commonly used neural network structures include feedback and feed-forward networks. The feedback networks are neural networks that have connections between network output and some or all other neuron units ［shown in Figure 5.15(a)］. Certain unit outputs in the figure are used as activated inputs to the network, and other unit outputs are used as network outputs. Due to the feedback, there is no guarantee that the networks become stable. To guarantee stability, constraints on synaptic weights are introduced so that the dynamics of the feedback network is expressed by a Lyapunov function. Concretely, a constraint of equivalent mutual connection weights of two units is implemented. The Hopfield network[4] is one such neural network.

Two aspects of a Hopfield network are as follows:
- Synaptic weights are determined by analytically solving the constraints, not by performing an iterative learning process. The weights are fixed during operation of the Hopfield network.
- Final network outputs are obtained by operating feedback networks, for the solution of an application task.

Another type of neural network which should be compared with the feedback network is the feed-forward type as shown in Figure 5.15(b). The feed-forward network is a filter whose output is the processed input signal. An algorithm will determine the synaptic weights to make the outputs match the desired result. These learning algorithms are categorized into supervised and unsupervised learning.

In a supervised learning algorithm, the synaptic weights are adjusted using input-output data, whereby input-output characteristics of the network are modified to obtain a desired output. In reinforcement learning, only a qualitative indicator of whether the desired output is achieved would be available.

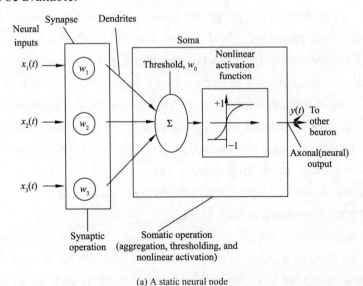

(a) A static neural node

Figure 5.15　Commonly used neural network structures

Note: $[x_1(t), \ldots, x_n(t)]$ represent neural inputs; $[w_1, \ldots, w_n]$ are the static synaptic weights; w_0 is the threshold; $y(t)$ is the axonal (neural) output

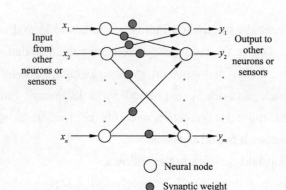

(b) A static (feed-forward) neural network with n-inputs and m-outputs

Figure 5.15   Commonly used neural network structures (continued)

In an unsupervised learning algorithm, the synaptic weights are adjusted according to the input values of the network, and not according to supervised output data of the network. Since the output characteristics are determined by the network itself, without the benefit of the desired output data, the unsupervised learning mechanism is called self-organization. Hebbian learning and competitive learning are representative of unsupervised learning algorithms.

(4) Fuzzy control + neural networks

Since intelligent control is a special class of highly nonlinear control, neural networks may be appropriately employed there, either separately or in conjunction with other techniques such as fuzzy control. Fuzzy-neural techniques are applicable in intelligent adaptive control, in particular, when parameter changes and unknown disturbances have to be compensated. The development of hybrid control systems by the fusion of fuzzy logic and neural networks is now becoming popular and holds much promise for the control of nonlinear, uncertain, and complex systems. Fuzzy logic control which is based on the concepts of fuzzy logic, first developed by Lotfi Zadeh, is a class of knowledge-based control that incorporates control knowledge in the form of a set of linguistic rules. This allows the application of heuristic information to be used in the control of complex systems.

Attempts have been made to combine the powerful learning and generalization capabilities of a neural network with the structure-rich characteristics of fuzzy logic control. The following is a brief summary of such cases:

1) Fuzzy rule generation using NNs

Given a series of sample data vectors, neural networks can be used to adaptively infer the fuzzy rule base using an unsupervised learning algorithm, such as an adaptive vector quantizer (AVQ) with differential competitive learning (DCL). This is useful when sample data can be relatively easily obtained. For example, in many industrial processes, a control system can be fine-tuned manually and the input-output data pairs can be recorded. When building the knowledge base, one can use these data vectors to infer the control surface using NNs. This can then be used either to replace the process of interviewing operators, or as a compliment to that process.

2) Use of NNs to tune fuzzy membership functions

In fuzzy control applications, often a fuzzy control system is developed by choosing fuzzy sets for the condition and action variables and building the knowledge base by interviewing operators. In this process, the selection of fuzzy sets is very subjective and standard rules to follow do not exist. With NNs, the fuzzy membership functions can be "optimized". This can be achieved via supervised learning. Given the training input data and the desired output value, NNs can be used to adjust the parameters of the membership functions to minimize the error between the desired output and the real output. As a result, the membership functions would be optimized for this particular application.

3) Use fuzzy rules to pretrain NNs

Once the fuzzy rules are known, they can be used to train a NN which can subsequently be used to replace a fuzzy tuner. This idea has been implemented to tune a servo-motor system using a NN trained by fuzzy rules. A similar idea may also be used to implement a fuzzy tuning algorithm for a hierarchical supervisory control system. Since a neural network can extrapolate and interpolate, reasonable tuning is maintained even when the conditions demanded by a rule are not exactly satisfied. Unlike the conventional approaches to fuzzy tuning, a NN-based fuzzy tuning method has a coupling effect on the input attributes, i.e., if two or more performance attributes are out of specification at one time, all of them will be taken into consideration in generating the output action attributes. Hence, less time would be required to tune the system.

The advanced/intelligent control techniques have nonlinear mapping capability, massive parallel hardware implementation potentials, and ability of learning and adapting under uncertain and noisy situations. These advantages make them desirable for a wide class of control applications, such as cloning experts, tracking trajectories or setpoints, and optimization (i.e. approximate dynamic programming). It is clear that neural networks offer a great potential to devise new and innovative control strategies for complex systems. It has also showed that as systems are becoming increasingly complex with the advancement of technology, hybrid systems which can combine the capabilities of other technology such as fuzzy control techniques with the learning capability of neural networks will prove to be more powerful. These technologies can help future control systems become autonomous, flexible, adaptable, and self-learning in highly dynamic environments.

Words and Expressions

control system [kənˈtrəʊl ˈsɪstəm]	控制系统
controller [kənˈtrəʊlə] someone who maintains and audits business accounts	n. 控制器
remote control [rɪˈməʊt kənˈtrəʊl]	远程控制
plant [plɑːnt] buildings for carrying on industrial labor	n. 工厂
open-loop system [ˈəʊpən luːp ˈsɪstəm]	开环系统
closed-loop system [kləʊzd luːp ˈsɪstəm]	闭环系统
stability [stəˈbɪlɪti] the quality or attribute of being firm and steadfast	n. 稳定性
transient response [ˈtrænziənt rɪˈspɒns]	瞬态响应
steady-state error [ˈstedi steɪt ˈerə(r)]	稳态误差
power amplification [ˈpaʊə ˌæmpləfɪˈkeɪʃn]	功率放大
gain [ɡeɪn] a quantity that is added	n. 增益
disturbance [dɪsˈtɜːbəns] activity that is an intrusion or interruption	n. 扰动
compensate [ˈkɒmpenseɪt] adjust for	v. 补偿
sensor [ˈsensə] any device that receives a signal or stimulus (as heat or pressure or light or motion etc.) and responds to it in a distinctive manner	n. 传感器
reference signal [ˈrefrəns ˈsɪɡnəl]	参考信号
robot arm [ˈrəʊbɒt ɑːm]	机械手
transducer [trænzˈdjuːsə(r)] an electrical device that converts one form of energy into another	n. 传感器；变频器；变换器
potentiometer [pəˌtenʃiˈɒmɪtə(r)] a measuring instrument for measuring direct current electromotive forces	n. 电位计；分压计；电势计；分压器
thermistor [θɜːˈmɪstə] a semiconductor device made of materials whose resistance varies as a function of temperature; can be used to compensate for temperature variation in other components of a circuit	n. 电热调节器

PID controller [kənˈtrəʊlə]	控制器
proportional control [prəˈpɔːʃənəl kənˈtrəʊl]	比例控制
integral control [ˈɪntɪgrəl kənˈtrəʊl]	积分控制
derivative control [dɪˈrɪvətɪv kənˈtrəʊl]	微分控制
optimal control [ˈɒptəməl kənˈtrəʊl]	最优控制
unstable plants [ʌnˈsteɪbəl pˈlɑːnts]	不稳定被控对象
DC motor [ˈməʊtə]	直流电机
adaptive control [əˈdæptɪv kənˈtrəʊl]	自适应控制
fuzzy logic control [ˈfʌzi ˈlɒdʒɪk kənˈtrəʊl]	模糊逻辑控制
neural network control [ˈnjʊərəl ˈnetwɜːk kənˈtrəʊl]	神经网络控制
model-reference adaptive system [ˈmɒdl ˈrefrəns əˈdæptɪv ˈsɪstəm]	模型参考自适应系统
artificial neural network [ˌɑːtɪˈfɪʃəl ˈnjʊərəl ˈnetwɜːk]	人工神经网络
fuzzify a variable [ˈfʌzɪfaɪ ə ˈveəriəbl]	模糊化过程
rule base [ruːl beɪs]	规则库
fuzzy inference [ˈfʌzi ˈɪnfərəns]	模糊推理
defuzzy [deˈfʌzi]	反模糊化
fuzzy sets [ˈfʌzi ˈsets]	模糊集
feedback network [ˈfiːdbæk ˈnetwɜːk]	反馈神经网络
feed-forward network [fiːd ˈfɔːwəd ˈnetwɜːk]	前馈神经网络
neuron [ˈnjʊərɒn] a cell that is specialized to conduct nerve impulses	n. 神经元
synaptic weight [sɪˈnæptɪk weɪt]	突触权值
supervised learning [ˈsjuːpəvaɪzd ˈlɜːnɪŋ]	监督学习
reinforcement learning [ˌriːɪnˈfɔːsmənt ˈlɜːnɪŋ]	强化学习
unsupervised learning algorithm [ˌʌnˈsjuːpəvaɪzd ˈlɜːnɪŋ ˈælɡərɪðəm]	无监督学习算法
hybrid system [ˈhaɪbrɪd ˈsɪstəm]	混杂系统

Notes

[1] VHDL 语言是一种用于电路设计的高级语言。它在 20 世纪 80 年代的后期出现。最初是由美国国防部开发出来供美军用来提高设计的可靠性和缩减开发周期的一种使用范围

较小的设计语言。VHDL 和 Verilog 作为 IEEE 的工业标准硬件描述语言，得到众多 EDA 公司支持，在电子工程领域，已成为事实上的通用硬件描述语言。

[2] FPGA 的开发相对于传统 PC、单片机的开发有很大不同。FPGA 以并行运算为主，以硬件描述语言来实现。以硬件描述语言（Verilog 或 VHDL）所完成的电路设计，可以经过简单的综合与布局，快速地烧录至 FPGA 上进行测试，是现代 IC 设计验证的技术主流。

[3] 通过模糊推理得到的结果是一个模糊集合或者隶属度函数，但在实际使用中，特别是在模糊控制中，必须要有一个确定的值才能去控制驱动或伺服机构。在推理得到的模糊集合中取一个能最佳代表这个模糊推理结果可能性的精确值的过程就称为反模糊化（又称精确化过程）。

[4] Hopfield 神经网络是一种具有相互连接的反馈型神经网络，具备联想记忆和快速优化能力，得到了广泛应用。

Questions for Discussion

1. Give an example of open-loop systems.
2. Name two applications for feedback control systems.
3. Functionally, how do closed-loop systems differ from open-loop systems?
4. What's the two major measures of performance of a control system?
5. What's the three major design criteria for control systems?
6. What's the characteristic and limits of proportional control?
7. What's the characteristic and limits of integral control?
8. What's the characteristic and limits of derivative control?
9. What's the limits of PID control algorithm?
10. Please name some advanced control methods.
11. What's the main steps for designing fuzzy logic controller?
12. How do we guarantee stability of neural networks?
13. What's the advantages of advanced/intelligent control?

5.2 On-Line Monitoring and Fault Diagnosis of Mechatronic Systems

5.2.1 Mechatronic Systems

Mechatronic Systems are research subjects of Mechatronics. Mechatronics is a multidisciplinary field of engineering that includes a combination of mechanical engineering, electronics, computer engineering, telecommunications engineering, systems engineering and control engineering. As technology advances, the subfields of engineering multiply and adapt. Mechatronics' aim is a design process that unifies these subfields. Originally, mechatronics just included the combination of mechanics and electronics, therefore the word is a combination of mechanics and electronics; however, as technical systems have become more and more complex the definition has been broadened to include more technical areas.

What is mechatronics

The word "mechatronics" originated in Japanese-English and was created by Tetsuro Mori, an engineer of Yaskawa Electric Corporation. The word "mechatronics" was registered as trademark by the company in Japan with the registration number of "46-32714" in 1971. However, afterward the company released the right of using the word to public, and the word "mechatronics" spread to the rest of the world. Nowadays, the word is translated in each language and the word is considered as an essential term for industry. Mechatronics has a bright future and is currently applied in everyday life for solutions ranging from transportation to optical telecommunication and biomedical engineering.

French standard NF E 01-010 gives the following definition: "approach aiming at the synergistic integration of mechanics, electronics, control theory, and computer science within product design and manufacturing, in order to improve and/or optimize its functionality". Figure 5.16 shows an overview of various fields that makes mechatronics. The control system is part of the loop for operating continuously in the surroundings. The mechanical systems contributes design, manufacturing and system dynamics. Computers contribute data acquisition method and algorithms. Electrical systems include DC and AC circuit analysis, power analysis and semiconductor device analysis.

Many people treat "mechatronics" as a modern buzzword synonymous with robotics (shown in Figure 5.17) and "electromechanical engineering".

Figure 5.16 An overview of various fields that makes mechatronics

Figure 5.17 Mechatronic systems: robotics

Mechatronics provides solutions that are efficient and reliable systems. Mechatronic

Systems mostly have microcomputers to ensure smooth functioning and higher dependability. The sensors in these systems absorb signals from the surroundings; react to these signals using appropriate processing to generate acquired output signals. Few examples of mechatronic systems are automated guided vehicles, robots, digitally controlled combust engines and machine tools with self-adaptive tools, aircraft flight control and navigation systems, and smart home appliances (i.e. washers, dryers, etc.).

5.2.2 On-Line Monitoring

Although monitoring has been around since the early 1960s with the advent of debuggers[1], the field has recently made some exciting advances. Monitoring systems today monitor distributed applications and are often themselves distributed[2]. In addition, they are increasingly seen as a viable solution to areas of growing concern: lack of dependability and tools to support distributed applications. Monitoring has succeeded in these areas and has matured in its ability to give users freedom in defining what is to be monitored.

Monitoring gathers information about a computational process as it executes and can be classified by its functionality (shown in Figure 5.18). Dependability includes fault tolerance and safety. Performance enhancement includes dynamic system configuration, dynamic program tuning, and on-line steering. Correctness checking is the monitoring of an application to ensure consistency with a formal specification. It can be used to detect runtime errors or as a verification technique. Security monitoring attempts to detect security violations such as illegal login or attempted file access. Control includes cases where the monitoring system is part of the target system, a necessary component in providing computational functionality. Debugging and testing employs monitoring techniques to extract data values from an application being tested. Performance evaluation uses monitoring to extract data from a system that is later analyzed to assess system performance.

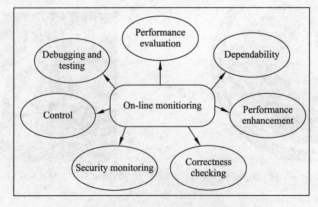

Figure 5.18   Primary uses of monitoring

Four of the seven functional areas are focused: dependability, performance enhancement, correctness checking, and security. The systems in these functional areas exhibit common characteristics. First, the monitor functions as an external observer of the target software. Unlike control monitors, external observers are not required to provide computational functionality. Second, the systems are designed to monitor the target software and respond while the target software is operational. This forces the monitoring system to react in a timely manner to events as they occur in the target system. Debuggers are not so constrained, because they either slow the application's execution rate or simply gather trace data for later analysis or replay. Lastly, the monitoring component is a permanent part of the overall system, although at times it may run at reduced functionality. This is unlike performance evaluation tools that are, like some hardware test tools, attached to a system.

We call a monitoring system that as follows:
- That is an external observer;
- That monitors a fully functioning application;
- That is generally intended to be permanent an on-line monitoring system.

These systems often do more than just gather information; they interpret the gathered information and respond appropriately. On-line monitoring systems can therefore provide increased robustness, security, fault-tolerance, and adaptability.

5.2.3 Fault Diagnosis

Fault diagnosis is also called fault detection, isolation, and recovery (FDIR)[3], which is a subfield of control engineering which concerns itself with monitoring a system, identifying when a fault has occurred, and pinpointing the type of fault and its location. Two approaches can be distinguished: ① A direct pattern recognition of sensor readings that indicate a fault and an analysis of the discrepancy between the sensor readings and expected values, derived from some model. ② In the latter case, it is typical that a fault is said to be detected if the discrepancy or residual goes above a certain threshold. It is then the task of fault isolation to categorize the type of fault and its location in the machinery. Fault detection and isolation (FDI) techniques can be broadly classified into two categories. These include model-based FDI and signal processing based FDI[4].

Fault diagnosis and small wind turbine

In model-based FDI techniques some model of the system is used to decide about the occurrence of fault. The system model may be mathematical or knowledge based. Some of the model-based FDI techniques include observer-based approach, parity-space approach, and parameter identification based methods. There is another trend of model-based FDI schemes, which is called set-membership methods. These methods guarantee the detection of fault under certain conditions. The main difference is that instead of finding the most likely model, these techniques omit the models, which are not compatible with data.

In signal processing based FDI, some mathematical or statistical operations are performed on the measurements, or some neural network is trained using measurements to extract the information about the fault.

Fault diagnosis of mechatronic systems is also called machine fault diagnosis, which is a field of mechanical engineering concerned with finding faults arising in machines. A particularly well developed part of it applies specifically to rotating machinery, one of the most common types encountered. To identify the most probable faults leading to failure, many methods are used for data collection, including vibration monitoring, thermal imaging, oil particle analysis, etc. Then these data are processed utilizing methods like spectral analysis, wavelet analysis, wavelet transform, short term Fourier transform, Gabor expansion, Wigner-Ville distribution (WVD), cepstrum, bispectrum, correlation method, high resolution spectral analysis, waveform analysis (in the time domain, because spectral analysis usually concerns only frequency distribution and not phase information) and others[5]. The results of this analysis are used in a root cause failure analysis in order to determine the original cause of the fault. For example, if a bearing fault is diagnosed, then it is likely that the bearing was not itself damaged at installation, but rather as the consequence of another installation error (i.e. misalignment) which then led to bearing damage. Diagnosing the bearing's damaged state is not enough for precision maintenance purposes. The root cause needs to be identified and remedied. If this is not done, the replacement bearing will soon wear out for the same reason and the machine will suffer more damage, remaining dangerous. Of course, the cause may also be visible as a result of the spectral analysis undertaken at the data-collection stage, but this may not always be the case.

The most common technique for detecting faults is the time-frequency analysis technique. For a rotating machine, the rotational speed of the machine (often known as the RPM), is not a constant, especially not during the startup and shutdown stages of the machine. Even if the machine is running in the steady state, the rotational speed will vary around a steady-state mean value, and this variation depends on load and other factors. Since sound and vibration signals obtained from a rotating machine which are strongly related to its rotational speed, it can be said that they are time-variant signals in nature. These time-variant features carry the machine fault signatures. Consequently, how these features are extracted and interpreted is important to research and industrial applications.

The most common method used in signal analysis is the FFT, or Fourier transform. The Fourier transform and its inverse counterpart offer two perspectives to study a signal: via the time domain or via the frequency domain. The FFT-based spectrum of a time signal shows us the existence of its frequency contents. By studying these and their magnitude or phase relations, we can obtain various types of information, such as harmonics, sidebands, beat frequency, bearing fault frequency and so on. However, the FFT is only suitable for signals whose frequency contents do not change over time; however, as mentioned above, the frequency contents of the sound and vibration signals obtained from a rotating machine are very much time-dependent. For this reason, FFT-based spectra are unable to detect how the frequency contents develop over time. To be more specific, if the RPM of a machine is increasing or decreasing during its startup or shutdown period, its bandwidth in the FFT spectrum will become much wider than it would be simply for the steady state. Hence, in such a case, the harmonics are not so distinguishable in the spectrum.

The time frequency approach for machine fault diagnosis can be divided into two broad categories: linear methods and the quadratic methods. The difference is that linear transforms can be inverted to construct the time signal, thus, they are more suitable for signal processing, such as noise reduction and time-varying filtering. Although the quadratic method describes the energy distribution of a signal in the joint time frequency domain, which is useful for analysis, classification, and detection of signal features, phase information is lost in the quadratic time-frequency representation; also, the time histories cannot be reconstructed with this method.

The short-term Fourier transform (STFT) and the Gabor transform are two algorithms commonly used as linear time-frequency methods. If we consider linear time-frequency analysis to be the evolution of the conventional FFT, then quadratic time frequency analysis would be the power spectrum counterpart. Quadratic algorithms include the Gabor spectrogram, Cohen's class and the adaptive spectrogram. The main advantage of time frequency analysis is discovering the patterns of frequency changes, which usually represent the nature of the signal. As long as this pattern is identified the machine fault associated with this pattern can be identified. Another important use of time frequency analysis is the ability to filter out a particular frequency component using a time-varying filter.

Words and Expressions

Mechatronics [mekə'trɒnɪks]	*n.* 机械电子学
on-line monitoring [ɒn'laɪn 'mɒnɪtərɪŋ]	在线监测
fault diagnosis [fɔːlt ˌdaɪəɡ'nəʊsɪs]	故障诊断
multidisciplinary [ˌmʌltidɪsə'plɪnəri] of or relating to the study of one topic, involving several subject disciplines	*adj.* 多学科的
multiply and adapt ['mʌltɪplaɪ ænd ə'dæpt]	交叉融合

buzzword synonymous [ˈbʌzwɜːd sɪˈnɒnəməs]	流行术语近义词
reliable [rɪˈlaɪəbl] worthy of reliance or trust	*adj.* 可靠的
dependability [dɪˌpendəˈbɪləti] the quality of being dependable or reliable	*n.* 可靠性
combust engine [kəmˈbʌst ˈendʒɪn]	燃油引擎
viable [ˈvaɪəbl] capable of being done with means at hand and circumstances as they are	*adj.* 切实可行的
functionality [fʌŋkʃəˈnæləti] the range of functions that a system can perform	*n.* 功能
fault tolerance [fɔːlt ˈtɒlərəns]	故障容错
consistency [kənˈsɪstənsi] the property of holding together and retaining its shape	*n.* 一致
tune [ˈtjuːn] to calibrate something (an instrument or electronic circuit) to a standard frequency	*v.* 调节
steer [ˈstɪrɪŋ] to guide or show the way	*v.* 引导
runtime error [rʌnˈtaɪm ˈerə(r)]	运行错误
operational [ɒpəˈreɪʃənl] pertaining to a process or series of actions for achieving a result	*adj.* 可操作的
trace data [treɪs ˈdeɪtə]	跟踪数据
permanent [ˈpɜːmənənt] a series of waves in the hair made by applying heat and chemicals	*adj.* 定常的
reduced functionality [rɪˈdjuːst fʌŋkʃəˈnæləti]	功能限制
robustness [rəʊˈbʌstnəs] the property of being strong and healthy in constitution	*n.* 鲁棒性
adaptability [əˌdæptəˈbɪləti] the ability to change (or be changed) to fit changed circumstances	*n.* 自适应性

pinpoint [ˈpɪnpɔɪnt] a very brief moment	n. 精准定位
pattern recognition [ˈpætən ˌrekəgˈnɪʃən]	模式识别
discrepancy [dɪsˈkrepənsi] a difference between conflicting facts or claims or opinions	n. 矛盾；异同
derive [dɪˈraɪv] forme or develop from something else; not original	v. 推导
residual [rɪˈzɪdjuəl] something left after other parts have been taken away	adj. 剩余的，这里指理想与实际的差值
parity-space [ˈpærɪtiː speɪs]	奇偶校验
set-membership [set ˈmembəʃɪp]	集合成员
compatible [kəmˈpætəbl] able to exist and perform in harmonious or agreeable combination	n. 兼容设备
encounter [ɪnˈkaʊntə(r)] to experience sth. unexpected	v. 遇到
thermal imaging [ˈθɜːməl ˈɪmɪdʒɪŋ]	热成像
root cause failure analysis [ruːt kɔːz ˈfeɪljə əˈnæləsɪs]	根因错误分析
bearing [ˈbeərɪŋ] relevant relation or interconnection	n. 轴承
misalignment [mɪsəˈlaɪnmənt] the spatial property of things that are not properly aligned	n. 未对准；角误差
remedy [ˈremədi] to correct an error or a fault or an evil	v. 纠正，改进
RPM (revolutions per minute) [revəˈluːʃəns pɜː ˈmɪnɪt]	转每分钟
steady-state mean value [ˈstedi steɪt miːn ˈvælju]	稳态均值
time-variant [taɪm ˈveəriənt]	时变
FFT (fast Fourier transformation) [fɑːst ˈfʊriə ˌtrænsfəˈmeɪʃən]	快速傅里叶变换
inverse counterpart [ɪnˈvɜːs ˈkaʊntəpɑːt]	反变换
harmonics [hɑːˈmɒnɪks] the study of musical sound	n. 谐波

sideband [ˈsaɪdbænd]	n. 边带
beat frequency [biːt ˈfriːkwənsi]	差频；拍频
time-dependent [taɪm dɪˈpendənt]	依赖时间的
distinguishable [dɪˈstɪŋgwɪʃəbl] capable of being perceived as different or distinct	adj. 可分辨的
quadratic method [kwɒˈdrætɪk ˈmeθəd]	二次方法
noise reduction [nɔɪz rɪˈdʌkʃən]	减噪
time-varying filtering [taɪm ˈveərɪŋ ˈfɪltərɪŋ]	时变滤波
joint time frequency domain [dʒɔɪnt taɪm ˈfriːkwənsi dəʊˈmeɪn]	联合时频域
reconstruct [ˌriːkənˈstrʌkt] adapt to social or economic change	v. 重构

Notes

[1] with the advent of debuggers 可译为"伴随调试器的出现"。

[2] 本句可译为：监测系统现在监测分布式应用，并且自身也是分布式的。

[3] fault detection, isolation, and recovery (FDIR) 可译为"故障检测，隔离和恢复"。

[4] 本句可译为：包含基于模型的和基于信号处理的故障检测隔离。

[5] 这些数据处理方法包含谱分析、小波分析、小波变换、短期傅里叶变换、Gabor 展开、Wigner-Ville 分布、倒频谱、双频谱、相关性、高精度谱分析、小波分析（时域的，因为谱分析通常只关心谱分布而不关心相位信息）及其他。

Questions for Discussion

1. What's your impression on Mechatronics? How many fields does Mechatronics include?
2. What's the functionalities of on-line monitoring?
3. How many classes does fault diagnosis contain?
4. Can you design an on-line monitoring and fault diagnosis system for a mechatronic system in your daily life?

5.3 Fluid Transmission and Control

5.3.1 Introduction of Fluid Transmission and Control

(1) Introduction of fluid transmission[1]

A fluid is defined as a substance that cannot sustain a shearing stress. A fluid can be liquid or gaseous. The science of fluid power is concerned with the utilization of pressurized liquid or gas to transmit power, but we will be dealing exclusively with hydraulic fluids. Utilization of fluid power is important because it is one of the three available means of transmitting power. Other methods of transmitting power are by utilizing mechanical means and by applying electrical energy. To demonstrate this we will consider that we have a prime mover such as a diesel engine on one side of the room and a mechanical contrivance on the other. The objective is to see how, in a generic sense, power can be used by the methods quoted above to perform the necessary mechanical work.

For mechanical power transmission, the prime mover is connected to the device and, by use of gearboxes, pulleys, belts and clutches, the necessary work can be performed. With the electrical method, an electrical generator is used. The current developed can be carried through electrical cable to operate electrical motors, linear or rotary, modulation being provided by variable resistance or solid state devices in the circuits. For fluid power utilization, an oil pump is connected to the engine and instead of electrical cables, high pressure hose is used to convey pressurized fluid to motors (again linear or rotary), pressure and flow modulation now being provided within the motors or by means of hydraulic valves. Any of the three methods described may be used however, if an engineering system requires: minimum weight and volume; large forces and low speeds; instant reversibility; remote control. Thus, the fluid power technique will often have significant competitive advantages.

Hydraulic control systems

1) Types of hydraulic fluids

Two basic types of hydraulic fluids used in control systems can be distinguished: petroleum base fluids and synthetic fluids. The synthetic fluids may be subdivided into chemically compounded and water base fluids. Petroleum base fluids are obtained from refining crude oil. The major disadvantages of these fluids are their potential fire hazard and restricted operational temperature range. To overcome these difficulties, synthetic fluids have been formulated from compounds which are chemically resistant to burning. Water, sometimes with soluble oil additives to increase lubricity and reduce rusting, is used in some industrial applications where large quantities of fluid are required and performance is not a premium. However, water is a poor hydraulic fluid because of its restrictive liquid range, low viscosity and lubricity, and rusting ability.

2) Selection of the hydraulic fluid

Many petroleum and synthetic fluids are available and more are being formulated. The highly technical formulations of the fluids with their various pros and cons makes the selection of such fluids difficult for those who are not thoroughly acquainted with the latest improvements and new formulations.

Generally, hydraulic fluids are chosen based on considerations of the environment of the application and chemical properties of the fluid. Physical properties such as viscosity, density, and bulk modulus are not usually basic considerations. Viscosity is very important, but usually a variety of viscosity characteristics are available in each fluid type. Bulk modulus should be large, but this requirement usually yields to the high temperature capability of the fluid. For example, the low bulk modulus of silicone fluids is more than offset by their high temperature range.

A basic judgment in fluid selection is required concerning the fire and explosion hazard posed by the application. If the environment and high temperature limit of the application are within the range of petroleum base fluids, then any number of suitable oils are available from numerous manufacturers. If the application requires a fire-resistant fluid, a choice must be made between the chemically compounded and water base synthetics. Factors to be considered are temperature range, cost, lubricity, compatibility, chemical, and handling characteristics of the fluid. Once a fluid type is selected, a number of viscosity and viscosity-temperature characteristics are usually made available, and a suitable matching must be made to the requirements of the system hardware. Consultation with representatives of hardware and fluid manufacturers is essential to ensure satisfactory compatibility and performance.

3) Advantages of using fluid power systems

It was stated earlier that there are advantages to using hydraulic systems rather than mechanical or electrical systems for specific applications and for those applications using large powers. Some of these advantages are given below:

- Force multiplication is possible by increasing actuator area or working pressure. In addition, torques and forces generated by actuators are limited only by pressure and as a result high power to weight ratio and high power to volume ratio are readily achievable.
- It is possible to have a quick acting system with large (constant) forces operating at low speeds and with virtually instant reversibility. In addition, a wide speed range of operating conditions may be achieved.
- A hydraulic system is relatively simple to construct with fewer moving parts than in comparable mechanical or electrical machines.
- Power transmission to remote locations is also possible provided that conductors and actuators can be installed at these locations.
- In most cases the hydraulic fluid circulated will act as a lubricant and will also carry away the heat generated by the system.
- A complex system may be constructed to perform a sequence of operations by means of mechanical devices such as cams, or electrical devices such as solenoids, limit

switches, or programmable electronic controls.

(2) The advantages and disadvantages of hydraulic control[2]

The increasing amount of power available to man that requires control and the stringent demands of modern control systems have focused attention on the theory, design, and application of control systems. Hydraulics—the science of liquid flow—is a very old discipline which has commanded new interest in recent years, especially in the area of hydraulic control, and fills a substantial portion of the field of control. Hydraulic control components and systems are found in many mobile, airborne, and stationary applications.

1) Advantages of hydraulic control

There are many unique features of hydraulic control compared to other types of control. These are fundamental and account for the wide use of hydraulic control. Some of the advantages are the following:

- Heat generated by internal losses is a basic limitation of any machine. Lubricants deteriorate, mechanical parts seize, and insulation breaks down as temperature increases. Hydraulic components are superior to others in this respect since the fluid carries away the heat generated to a convenient heat exchanger. This feature permits smaller and lighter components. Hydraulic pumps and motors are currently available with horsepower to weight ratios greater than 2 hp/lb. Small compact systems are attractive in mobile and airborne installations.
- The hydraulic fluid also acts as a lubricant and makes possible long component life.
- There is no phenomenon in hydraulic components comparable to the saturation and losses in magnetic materials of electrical machines. The torque developed by an electric motor is proportional to current and is limited by magnetic saturation. The torque developed by hydraulic actuators (i.e. motors and pistons) is proportional to pressure difference and is limited only by safe stress levels. Therefore hydraulic actuators develop relatively large torques for comparatively small devices.
- Electrical motors are basically a simple lag device from applied voltage to speed. Hydraulic actuators are basically a quadratic resonance from flow to speed with a high natural frequency. Therefore hydraulic actuators have a higher speed of response with fast starts, stops, and speed reversals possible. Torque to inertia ratios are large with resulting high acceleration capability. On the whole, higher loop gains and bandwidths are possible with hydraulic actuators in servo loops.
- Hydraulic actuators may be operated under continuous, intermittent, reversing, and stalled conditions without damage. With relief valve protection, hydraulic actuators may be used for dynamic breaking. Larger speed ranges are possible with hydraulic

actuators. Both linear and rotary actuators are available and add to the flexibility of hydraulic power elements.

- Hydraulic actuators have higher stiffness, that is, inverse of slope of speed-torque curves, compared to other drive devices since leakages are low. Hence there is little drop in speed as loads are applied. In closed loop systems, this results in greater positional stiffness and less position error.
- Open and closed loop control of hydraulic actuators is relatively simple using valves and pumps.
- Other aspects compare less favorably with those of electromechanical control components but are not so serious that they deter wide use and acceptance of hydraulic control. The transmission of power is moderately easy with hydraulic lines, Energy storage is relatively simple with accumulators.

2) Disadvantages of hydraulic control

Although hydraulic controls offer many distinct advantages, several disadvantages tend to limit their use. Major disadvantages are the following:

- Hydraulic power is not so readily available as that of electrical power. This is not a serious threat to mobile and airborne applications but most certainly affects stationary applications.
- Small allowable tolerances results in high costs of hydraulic components.
- The hydraulic fluid imposes an upper temperature limit. Fire and explosion hazards exist if a hydraulic system is used near a source of ignition. However, these situations have improved with the availability of high temperature and fire resistant fluids. Hydraulic systems are messy because it is difficult to maintain a system free from leaks, and there is always the possibility of complete loss of fluid if a break in the system occurs.
- It is impossible to maintain the fluid free of dirt and contamination. Contaminated oil can clog valves and actuators and, if the contaminant is abrasive, cause a permanent in performance and/or failure. Contaminated oil is the chief source of hydraulic control failures. Clean oil and reliability are synonymous terms in hydraulic control.
- Basic design procedures are lacking and difficult to obtain because of the complexity of hydraulic control analysis. For example, the current flow through a resistor is described by a simple law—Ohm's law. In contrast, no single law exists which describes the hydraulic resistance of passages to flow. For this seemingly simple problem there are almost endless details of Reynolds number, laminar or turbulent flow, passage geometry, friction factors, and discharge coefficients to cope with. This factor limits the degree of sophistication of hydraulic control devices.

- Hydraulics are not so flexible, linear, accurate, and inexpensive as electronic or electromechanical devices in the manipulation of low power signals for purposes of mathematical computation, error detection, amplification, instrumentation, and compensation. Therefore, hydraulic devices are generally not desirable in the low power portions of control systems.

(3) Hydraulic control circuits

The outstanding characteristics of hydraulic power elements have combined with their comparative inflexibility at low power levels to make hydraulic controls attractive primarily in power portions of circuits (shown in Figure 5.19) and systems. The low power portions of systems are usually accomplished by mechanical or electromechanical means.

1) Open-loop

Figure 5.19(a) shows the open loop hydraulic circuit. Pump-inlet and motor-return (via the directional valve) are connected to the hydraulic tank. The term loop applies to feedback; the more correct term is open versus closed "circuit". Open center circuits use pumps which supply a continuous flow. The flow is returned to the tank through the control valve's open center; that is, when the control valve is centered, it provides an open return path to the tank and the fluid is not pumped to a high pressure. Otherwise, if the control valve is actuated it routes fluid to and from an actuator and tank. The fluid's pressure will rise to meet any resistance, since the pump has a constant output. If the pressure rises too high, fluid returns to the tank through a pressure relief valve. Multiple control valves may be stacked in series. This type of circuit can use inexpensive, constant displacement pumps.

2) Closed-loop

As can be seen from Figure 5.19(b), in a closed-loop hydraulic circuit, motor-return is connected directly to the pump-inlet. To keep up pressure on the low pressure side, the circuits have a charge pump (a small gear pump) that supplies cooled and filtered oil to the low pressure side. Closed-loop circuits are generally used for hydrostatic transmissions in mobile applications.

- Advantages: No directional valve and better response, the circuit can work with higher pressure. The pump swivel angle covers both positive and negative flow direction.
- Disadvantages: The pump cannot be utilized for any other hydraulic function in an easy way and cooling can be a problem due to limited exchange of oil flow.

High power closed loop systems generally must have a "flush-valve" assembled in the circuit in order to exchange much more flow than the basic leakage flow from the pump and the motor, for increased cooling and filtering. The flush valve is normally integrated in the motor housing to get a cooling effect for the oil that is rotating in the motor housing itself. The losses in the motor housing from rotating effects and losses in the ball bearings can be considerable as motor speeds will reach 4,000~5,000 rev/min or even more at maximum vehicle speed. The leakage flow as well as the extra flush flow must be supplied by the charge pump. A large charge pump is thus very important if the transmission is designed for high pressures and high motor speeds. High oil temperature is usually a major problem when using hydrostatic transmissions at high vehicle speeds for longer periods, for instance when transporting the machine from one work place to the other. High oil temperatures for long periods will drastically reduce the lifetime of the transmission. To keep down the oil temperature, the system pressure during transport must be lowered, meaning that the minimum displacement for the motor must be limited to a reasonable value. Circuit pressure during transport around 200~250 bar is recommended.

Closed loop systems in mobile equipment are generally used for the transmission as an alternative to mechanical and hydrodynamic (converter) transmissions. The advantage is a stepless gear ratio (continuously variable speed/torque) and a more flexible control of the gear ratio depending on the load and operating conditions. The hydrostatic transmission is generally limited to around 200 kW maximum power, as the total cost gets too high at higher power compared to a hydrodynamic transmission. Large wheel loaders for instance and heavy machines are therefore usually equipped with converter transmissions. Recent technical achievements for the converter transmissions have improved the efficiency and developments in the software have also improved the characteristics, for example selectable gear shifting programs during operation and more gear steps, giving them characteristics close to the hydrostatic transmission.

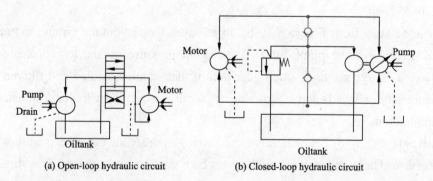

(a) Open-loop hydraulic circuit   (b) Closed-loop hydraulic circuit

Figure 5.19   Principal circuit diagram for open loop and closed loop system

Hydrostatic transmissions for earth moving machines, such as for track loaders, are often equipped with a separate "inch pedal" that is used to temporarily increase the diesel engine RPM while reducing the vehicle speed in order to increase the available hydraulic power output for the working hydraulics at low speeds and increase the tractive effort. The function is similar to stalling a converter gearbox at high engine RPM. The inch function affects the preset characteristics for the "hydrostatic" gear ratio versus diesel engine RPM.

(4) Hydraulic control systems

Hydraulic actuation devices may be controlled by a pump or a valve giving four basic hydraulic power elements and two basic overall systems: pump controlled and valve controlled.

1) Pump controlled system

The pump controlled system consists of a variable delivery pump supplying fluid to an actuation device. The fluid flow is controlled by the stroke of the pump to vary output speed, and the pressure generated matches the load. It is usually difficult to close couple the pump to the actuator and this causes large contained volumes and slow response. The main features of pump controlled systems are as following:

- Slow response because pressures must be built up, contained volumes are large, find the stroke servo has comparatively slow response.
- Much more efficient since both pressure and flow are closely matched to load requirements.
- Bulky power element size makes application difficult if pump is close coupled to actuator.
- Auxiliary pump and valving required to provide oil for replenishing and cooling.
- An electrohydraulic servo is generally required to stroke (lie pump which increases system cost and complexity.

2) Valve controlled system

The valve controlled system consists of a servo valve controlling the flow from a hydraulic power supply to an actuation device. The hydraulic power supply is usually a constant pressure type (as opposed to constant flow) and there are two basic configurations. One consists of a constant delivery pump with a relief valve to regulate pressure, whereas the other is much more efficient because it uses a variable delivery pump with a stroke control to regulate pressure. The main features of valve controlled systems are as following:

- Fast response to valve and load inputs because contained volumes are small and supply pressure is constant.
- Less efficient because supply pressure is constant regardless of load, and leakages are greater.
- Small and light power element but a bulky hydraulic power supply is required.
- Oil temperature builds up because of inefficiency which necessitates heat exchangers.
- Several valve-controlled systems can be fed from a single hydraulic power supply.

The features of each system tend to complement the other so that application requirements would dictate the choice to be made. Generally there is not a cost advantage to either because the need for a replenishing arrangement and a stroke servo for the pump controlled system offsets the costly servo valve and heat exchangers required for the valve controlled system. However, the faster response capability of valve controlled systems—both to valve and load inputs—makes this arrangement preferred in the majority of applications in spite of its lower theoretical maximum operating efficiency of 67%. In low power applications where the inefficiency is comparatively less important, use of valve controlled systems is nearly universal. Applications which require large horsepowers for control purposes usually do not require fast response so that a pump controlled system is preferred because of its superior theoretical maximum operating efficiency of 100%.

(5) Hydraulic components

1) Hydraulic pump

Hydraulic pumps supply fluid to the components in the system. Pressure in the system develops in reaction to the load. Hence, a pump rated for 5,000 psi is capable of maintaining flow against a load of 5,000 psi. Pumps have a power density about ten times greater than an electric motor (by volume). They are powered by an electric motor or an engine, connected through gears, belts, or a flexible elastomeric coupling to reduce vibration.

Hydraulic pumps are used in hydraulic drive systems and can be hydrostatic or hydrodynamic. A hydraulic pump is a mechanical source of power that converts mechanical power into hydraulic energy (i.e. flow, pressure). It generates flow with enough power to overcome pressure induced by the load at the pump outlet. When a hydraulic pump operates, it creates a vacuum at the pump inlet, which forces liquid from the reservoir into the inlet line to the pump and by mechanical action delivers this liquid to the pump outlet and forces it into the hydraulic system. Hydrostatic pumps are positive displacement pumps while hydrodynamic pumps can be fixed displacement pumps, in which the displacement (flow through the pump per rotation of the pump) cannot be adjusted, or variable displacement pumps, which have a more complicated construction that allows the displacement to be adjusted. Although, hydrodynamic pumps are more frequent in day-to-day life. Hydrostatics pump which are of various types works on the principle of Pascal's law. It states that the increase in pressure at one point of the enclosed liquid in equilibrium of rest is transmitted equally to all other points of the liquid, unless the effect of gravity is neglected.

In the positive displacement machine fluid passes through the inlet into a chamber which expands in volume and fills with fluid. The volume expansion causes shaft rotation a motor in contrast to a pump where the volume expansion is caused by shaft rotation. The volume of trapped fluid is then scaled from the inlet by some mechanical means and then transported to the outlet side where it is discharged. A succession of small volumes of fluid transported in this manner gives a fairly uniform flew. Thus a positive or definite amount of fluid is displaced or transported through the machine per unit of shaft revolution. Positive displacement machines are quite efficient and find extensive use in control systems.

Most pumps are working in open systems. The pump draws oil from a reservoir at atmospheric pressure. It is very important that there is no cavitation at the suction side of the pump. For this reason the connection of the suction side of the pump is larger in diameter than the connection of the pressure side. In case of the use of multi-pump assemblies, the suction connection of the pump is often combined. It is preferred to have free flow to the pump (pressure at inlet of pump at least 0.8 bar). The body of the pump is often in open connection with the suction side of the pump.

In case of a closed system, both sides of the pump can be at high pressure. The reservoir is often pressurized with 6~20 bars boost pressure. For closed loop systems, normally axial piston pumps are used. Because both sides are pressurized, the body of the pump needs a separate leakage connection.

Common types of hydraulic pumps to hydraulic machinery applications are as following:
- Gear pump: cheap, durable (especially in g-rotor form), simple. Less efficient, because they are constant (fixed) displacement, and mainly suitable for pressures below 20 MPa (3,000 psi) (shown in Figure 5.20).
- Vane pump: cheap and simple, reliable. Good for higher-flow low-pressure output.
- Axial piston pump: many designed with a variable displacement mechanism, to vary output flow for automatic control of pressure (shown in Figure 5.21). There are various axial piston pump designs, including swashplate and checkball. The most common is the swashplate pump. A variable-angle swashplate causes the pistons to reciprocate a greater or lesser distance per rotation, allowing output flow rate and pressure to be varied (greater displacement angle causes higher flow rate, lower pressure, and vice versa).

Figure 5.20　Gear pump　　　　　　　Figure 5.21　Axial piston pump

- Radial piston pump: normally used for very high pressure at small flows.
- Piston pumps: more expensive than gear or vane pumps, but provide longer life operating at higher pressure, with difficult fluids and longer continuous duty cycles.

Piston pumps make up one half of a hydrostatic transmission.

2) Control valves

All fluid power circuits incorporate valves. Hydraulic control valves are devices that use mechanical motion to control a source of fluid power. They vary in arrangement and complexity, depending upon their function. Because control valves are the mechanical (or electrical) to fluid interface in hydraulic systems, their performance is under scrutiny, especially when system difficulties occur. Therefore knowledge of the performance characteristics of valves is essential. The purpose of this chapter is to discuss the characteristic and design criteria for the principal types of hydraulic control valves. Although emphasis placed on valves for servo control, the principles involved apply equally well to valves used in other applications, such as solenoid valves, pressure reducing valves, and flow control valves.

There are three categories of valves, i.e., directional control, pressure control, and flow control. A simple application of a directional control valve would be the valve controlled manually by an operator that determines which end of a cylinder is connected to a pump. The most commonly encountered pressure control valve is the pressure relief valve used to protect components from excess forces caused by overloads or actuators reaching the end of their travel. A flow control valve is used to route oil to a secondary circuit in such a fashion that flow rate remains approximately constant even when pressure is varying.

The principle of operation of most valves is the same. A valve is a variable area orifice where the orifice area may be controlled by conditions in a circuit, for example a pressure relief valve operates without operator intervention. Alternatively the orifice area may be controlled by an operator as in a directional control valve. This categorization is a little simplistic because not all directional control valves are directly linked to an operator. Valves may be moved by electrical actuators and by pressure actuators. Thus many valves are quite complicated in terms of the number of parts in the valve. Valves may also have feedback loops within them and the stability may have to be examined using control theory. Figure 5.22 shows valves made by Bosch Rexroth, which is one of the leading companies in drive and control technologies.

Figure 5.22　Rexroth valves

Directional control valves route the fluid to the desired actuator. They usually consist of a spool inside a cast iron or steel housing. The spool slides to different positions in the housing, and intersecting grooves and channels route the fluid based on the spool's position.

The spool has a central (neutral) position maintained with springs; in this position the supply fluid is blocked, or returned to tank. Sliding the spool to one side routes the hydraulic fluid to an actuator and provides a return path from the actuator to tank. When the spool is moved to the opposite direction the supply and return paths are switched. When the spool is allowed to return to neutral (center) position the actuator fluid paths are blocked, locking it in position.

Directional control valves are usually designed to be stackable, with one valve for each hydraulic cylinder, and one fluid input supplying all the valves in the stack.

Tolerances are very tight in order to handle the high pressure and avoid leaking, spools typically have a clearance with the housing of less than a thousandth of an inch (25 μm). The valve block will be mounted to the machine's frame with a three point pattern to avoid distorting the valve block and jamming the valve's sensitive components.

The spool position may be actuated by mechanical levers, hydraulic pilot pressure, or solenoids which push the spool left or right. A seal allows part of the spool to protrude outside the housing, where it is accessible to the actuator.

The main valve block is usually a stack of off the shelf directional control valves chosen by flow capacity and performance. Some valves are designed to be proportional (flow rate proportional to valve position), while others may be simply on-off. The control valve is one of the most expensive and sensitive parts of a hydraulic circuit.

3) Hydraulic cylinder

A hydraulic cylinder (also called a linear hydraulic motor) is a mechanical actuator that is used to give a unidirectional force through a unidirectional stroke. It has many applications, notably in construction equipment, manufacturing machinery, and civil engineering.

Hydraulic cylinders get their power from pressurized hydraulic fluid, which is typically oil. The hydraulic cylinder consists of a cylinder barrel, in which a piston connected to a piston rod moves back and forth. The barrel is closed on one end by the cylinder bottom (also called the cap), and the other end by the cylinder head (also called the gland) where the piston rod comes out of the cylinder. The piston has sliding rings and seals. The piston divides the inside of the cylinder into two chambers, the bottom chamber (cap end) and the piston rod side chamber (rod end / head end). Figure 5.23 illustrates the internal components of a hydraulic cylinder.

Figure 5.23　A cut away of a welded body hydraulic cylinder showing the internal components

4) Reservoir

The hydraulic fluid reservoir holds excess hydraulic fluid to accommodate volume changes from: cylinder extension and contraction, temperature driven expansion and contraction, and leaks. The reservoir is also designed to aid in separation of air from the fluid and also work as a heat accumulator to cover losses in the system when peak power is used. Design engineers are always pressured to reduce the size of hydraulic reservoirs, while equipment operators always appreciate larger reservoirs. Reservoirs can also help separate dirt and other particulate from the oil, as the particulate will generally settle to the bottom of the tank. Some designs include dynamic flow channels on the fluid's return path that allow for a smaller reservoir.

5) Accumulators

A hydraulic accumulator is essentially a type of energy storage device, a pressure storage reservoir in which a non-compressible hydraulic fluid is held under pressure by an external source (shown in Figure 5.24). The external source can be a spring, a raised weight, or a compressed gas. A hydraulic accumulator enables a hydraulic system to: cope with extremes of demand using a less powerful pump; store power for intermittent duty cycles; provide emergency or standby power; respond more quickly to a temporary demand; smooth out pulsations; compensate for leakage loss.

Figure 5.24   Accumulators

6) Hydraulic fluid

Also known as tractor fluid, hydraulic fluid is the life of the hydraulic circuit. It is usually petroleum oil with various additives. Some hydraulic machines require fire resistant fluids, depending on their applications. In some factories where food is prepared, either an edible oil or water is used as a working fluid for health and safety reasons.

In addition to transferring energy, hydraulic fluid needs to lubricate components, suspend contaminants and metal filings for transport to the filter, and to function well to several hundred degrees Fahrenheit or Celsius.

7) Filters

Filters are an important part of hydraulic systems which removes the unwanted particles from fluid. Metal particles are continually produced by mechanical components and need to be removed along with other contaminants.

Filters may be positioned in many locations. The filter may be located between the reservoir and the pump intake. Blockage of the filter will cause cavitation and possibly failure of the pump. Sometimes the filter is located between the pump and the control valves. This arrangement is more expensive, since the filter housing is pressurized, but eliminates cavitation problems and protects the control valve from pump failures. The third common filter location is just before the return line enters the reservoir. This location is relatively insensitive to blockage and does not require a pressurized housing, but contaminants that enter the reservoir from external sources are not filtered until passing through the system at least once. Filters are used from 7 micron to 15 micron depends upon the viscosity grade of hydraulic oil.

8) Tubes, pipes and hoses

Hydraulic tubes are seamless steel precision pipes, specially manufactured for hydraulics. The tubes have standard sizes for different pressure ranges, with standard diameters up to 100 mm. The tubes are supplied by manufacturers in lengths of 6 m, cleaned, oiled and plugged. The tubes are interconnected by different types of flanges (especially for the larger sizes and pressures), welding cones/nipples (with O-ring seal), several types of flare connection and by cut-rings. In larger sizes, hydraulic pipes are used. Direct joining of tubes by welding is not acceptable since the interior cannot be inspected.

Hydraulic pipe is used in case standard hydraulic tubes are not available. Generally these are used for low pressure. They can be connected by threaded connections, but usually by welds. Because of the larger diameters the pipe can usually be inspected internally after welding. Black pipe is non-galvanized and suitable for welding.

Hydraulic hose is graded by pressure, temperature, and fluid compatibility. Hoses are used when pipes or tubes can not be used, usually to provide flexibility for machine operation or maintenance. The hose is built up with rubber and steel layers. A rubber interior is surrounded by multiple layers of woven wire and rubber. The exterior is designed for abrasion resistance. The bend radius of hydraulic hose is carefully designed into the machine, since hose failures can be deadly, and violating the hose's minimum bend radius will cause failure. Hydraulic hoses generally have steel fittings swaged on the ends. The weakest part of the high pressure hose is the connection of the hose to the fitting. Another disadvantage of hoses is the shorter life of rubber which requires periodic replacement, usually at five to seven year intervals.

Tubes and pipes for hydraulic applications are internally oiled before the system is commissioned. Usually steel piping is painted outside. Where flare and other couplings are used, the paint is removed under the nut, and is a location where corrosion can begin. For this reason, in marine applications most piping is stainless steel.

9) Seals, fittings and connections

Components of a hydraulic system [sources (i.e. pumps), controls (i.e. valves) and actuators (i.e. cylinders)] need connections that will contain and direct the hydraulic fluid without leaking or losing the pressure that makes them work. In some cases, the components can be made to bolt together with fluid paths built-in. In more cases, though, rigid tubing or flexible hoses are used to direct the flow from one component to the next. Each component has entry and exit points for the fluid involved (called ports) sized according to how much fluid is expected to pass through it.

There are a number of standardized methods in use to attach the hose or tube to the component. Some are intended for ease of use and service, others are better for higher system pressures or control of leakage. The most common method, in general, is to provide in each component a female-threaded port, on each hose or tube a female-threaded captive nut, and use a separate adapter fitting with matching male threads to connect the two. This is functional, economical to manufacture, and easy to service.

A typical piece of machinery or heavy equipment may have thousands of sealed connection points and several different types:

- Pipe fittings: the fitting is screwed in until tight, difficult to orient an angled fitting correctly without over or under tightening.
- O-ring boss: the fitting is screwed into a boss and orientated as needed, an additional nut tightens the fitting, washer and O-ring in place.
- Flare fittings, are metal to metal compression seals deformed with a cone nut and pressed into a flare mating.
- Face seal, metal flanges with a groove and O-ring seal are fastened together.
- Beam seals are costly metal to metal seals used primarily in aircraft.
- Swaged seals, tubes are connected with fittings that are swaged permanently in place.

5.3.2 Electro-Hydrostatic Actuators

Electro-hydrostatic actuators

Electro-hydrostatic actuators (EHA)[3], often referred to as "power by wire", are fully self-contained actuation systems that combine design elements from electric and electrohydraulic actuation. They receive power from an electric source and transform an input command signal (usually electrical) into motion.

More electric aircraft initiative has led to fly-by-wire (FBW) and power-by-wire (PBW) technologies. PBW technology has several advantages and has potential for weight reduction and energy saving compared to conventional hydraulic system. Recent development in aviation technology have combined the electric and hydraulic system and arrived at high performance electro hydrostatic actuator (EHA) system. It is a one of the PBW technology which aims at replacing the centralized hydraulic system by a local self-contained and compact actuator system. It is an emerging technology which combines the benefits of conventional hydraulic system and direct drive actuators, like the high energy efficiency, high dynamic response, high torque to mass ratio and high maintainability. EHA system was initially introduced in aircraft for flight control system.

EHA technology is being used as the basis of the revolutionary electric flight control system on Lockheed Martin's F-35 Lightning II and is adapted to provide thrust vector control on NASA's 2nd Generation Reusable Launch Vehicle program—Systems Research

Aircraft (shown in Figure 5.25). For industrial applications, it provides high force control with all electric input power.

The earlier flight control system was operated by mechanical systems which are heavy and contribute to lot of fuel consumption. But recent advancement has led to fly-by-wire and power-by-wire technology. In spite of many advantages of FBW system, the disadvantage is

Figure 5.25　Systems Research Aircraft (SRA)

that if a single actuator fails then the entire hydraulic system has to be stopped. Hence to avoid this consequence PBW technology came into existence. It completely replaces the centralized hydraulic system by a localized compact actuator system. The benefits of FBW are retained as it is and additional advantages are added by PBW technology to make the system more efficient and reliable. The need of PBW actuation system for the next generation all-electric aircraft concept has been explained. EHA is a one of the power-by-wire technology used to actuate the flight control surfaces in aircraft. Electro hydrostatic actuator is a self-contained, modular and compact actuator system. It aims at replacing a centralized hydraulic system by a localized, self-contained and compact actuator system. Figure 5.26 shows the block diagram of EHA system. It mainly consists of controller, servo motor[4], pump, control elements and actuator (rotary / linear). The servo motor driven bi-directional pump in turn supplies oil to the actuator. The speed and direction of actuator is controlled based on speed and direction of servo motor. Hence the need of servo valve is eliminated. Actuator may be rotary or linear type depending upon the requirement. The controller gives output signal based on error between desired displacement and measured displacement. The development of a suitable controller is a prime importance in EHA system for better dynamic performance.

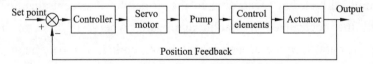

Figure 5.26　Block diagram of EHA

An integrated EHA combines high power density and integrates all the traditional function components—electric motor, reversing pump, special designed valves, reservoir, and double-acting hydraulic cylinder or hydraulic motor with light weight, low noise and a small envelope. Figure 5.27 shows a prototype of an EHA designed by Parker.

Compact EHA is designed to make commissioning as simple as possible. The motor is connected to a suitable power supply and switching circuit, and the rod or base end is secured with a pivot pin. The unit is then actuated to align the opposite pivot pin connection, and the pin inserted to secure. And that's it—your Compact EHA is ready for use. Because the compact EHA is flushed, filled and sealed for life, there is virtually no maintenance required. This, in combination with the anodized housing, stainless steel rod and rugged seals and components, provides a longer service life with reduced warranty costs.

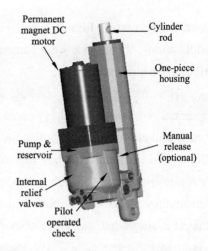

Figure 5.27　Prototype of a Parker EHA

Although electric motors are reliable, compact and controllable, they suffer from deficiencies which render them incomplete despite the fact that they are the most frequently used actuators for robotic applications. Owing to their favorable features, hybrid electro-hydrostatic actuators can compensate for those deficiencies. Application of EHA eliminate two important deficiencies: backlash (which are the result of tolerances of nominal dimensions and in-service wear of components used for power transmission—gear trains) and the small power-to-mass ratio.

EHA unifies a number of characteristics which meet requirements of modern robotic systems. According to laboratory tests reported by previously mentioned authors, EHA/IEHA successfully meets requirements such as: accuracy, precision, simple control, ability to integrate, modularity, flexibility, compactness, energy efficiency, small mass, continuous control of rotation direction under high payloads, etc.

The concept of EHA application requires that every joint or movable component has its own self-contained drive unit, which allows its independent operation even in case when some of the joints' drives fail (according to the principle of decentralization). This is not achievable with the centralized hydraulic aggregate, where, in case of its failure, the hydraulic system is out of service. Decentralization principle is also beneficial to maintenance. It is simpler and faster to replace a compact device installed within a joint, than to replace a complete central aggregate. One of the main drawbacks of the conventional hydraulic systems is that they rely on the central hydraulic aggregate. The central aggregate consists of one or more hydraulic pumps, which are actuated by one or several electric motors (or internal combustion engine, in case of a mobile application), and a spacious reservoir filled with hydraulic oil. The pump, electric motor and reservoir must be sized to meet requirements of all the actuators installed within robotic application.

The drawback of hydraulics is variable pressure, which significantly affects accuracy and precision of robot operation. In addition to positioning accuracy, this phenomenon can affect the durability of individual components, i.e., hydraulic hoses. Pressure oscillations can lead to premature failure of hydraulic hoses. Such fluctuations can never be entirely eliminated, but certain contemporary solutions can greatly alleviate the problem. Available solutions include application of accumulator, Helmholz resonator, and adaptive hydro-pneumatic pulsation damper.

In comparison to conventional hydraulic systems, the advantage of EHA device lies in the fact that it is fixed on an actuator (i.e. joint), minimizing the unnecessary losses in fluid energy transfer through pipe lines and hose lines. Moreover, EHA devices prevent potential hose line entanglement hazard during robot operation, which could otherwise lead to reduced hose cross section, resulting in pressure bursts and hydraulic surges. The use of EHA also increases system's dynamic ability and reduces the number of potential locations where the external oil leaks are likely to appear. The application of EHA in conjunction with symmetrical hydraulic cylinders and a bidirectional hydraulic motor, eliminates the need for flow valve or distributor valves, since flow direction and pump capacity are regulated by the integrated pump micro-valve. Another advantage of this device is the compact assembly of miniaturized hydraulic components with small overall dimensions and a very good ability of integration. Reduced overall dimensions contribute to space efficiency and lower mass of the resulting system in which EHA is installed.

Power transmission in a hydrostatic system can be realized in two ways: by using distribution valves or a variable displacement pump. Control by distribution valves is often used in industry but has limited flexibility. There is also some drawbacks while using single and double-acting cylinders. The difference in piston surface areas which are under pressure, results in the difference in piston rod forces and extension or retraction speeds. This problem can be partially solved using a differential distribution valve, whose ratio between the cross section areas of branches is proportional to the cross section areas ratio of piston with the single-action piston rod.

EHA system was initially introduced in aircraft for flight control system and then gradually it is finding many applications in industries and robotics. There are basically three configurations of the EHA system available. They are EHA system with variable pump and fixed motor (VPFM) [shown in Figure 5.28(a)], fixed pump and variable motor (FPVM) [shown in Figure 5.28(b)] and variable pump and variable motor (VPVM) [shown in Figure 5.28(c)]. Work on control strategies for EHA has highlighted the relative merits of the different configurations. The FPVM configuration of EHA is considered to have simpler structure and higher efficiency than EHA-VPFM. The EHA-VPFM has faster dynamic response than EHA-FPVM, but the efficiency is too low. In EHA-PVM mode, the displacement and rotating speed of pump can be adjusted simultaneously, therefore, it can combine the advantages of the other two types of EHA. The performance of closed loop EHA with single rod cylinder load operation system has been analyzed.

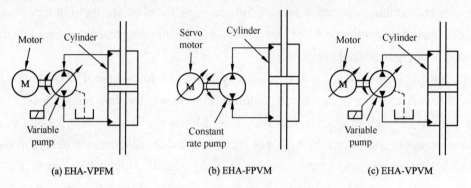

(a) EHA-VPFM   (b) EHA-FPVM   (c) EHA-VPVM

Figure 5.28   EHA systems

  Power-by-wire technology was developed for aerospace industry and further refined and improved to adapt EHA devices for industrial applications in robotics. The integration of hydraulic components which work with electrical components that control the hydraulic system, resulted in a number of advantages. Some of these were related to improved flexibility during operation, decreased energy use, smaller size, higher power-to-mass ratio and improved continuous control of operating parameters under high external payloads. Significant technological advances in manufacture of hydraulic components along with improved integration of embedded systems, lead to new applications of hydraulics in robotics. New materials and new manufacturing technologies allowed manufacture of miniature hydraulic components, thus saving space and decreasing loads on the structure, while designed to sustain large enough forces to perform tasks. Beside their tighter tolerances, the miniature components also had a minimum oil leakage, provided that assembly was well completed and that maintenance was timely.

  Hydraulic system designs, similar to EHA, are used in large and massive robots. One such application is the vehicle with adaptive suspension, which is the result of a military scientific project conducted at the University of Ohio during the eighties of the twentieth century. Sometime later, Bobrow published his paper on the development of a hydraulic drive for robotic applications, attempting to solve the problem which occurs when trying to control high torques using a combination of electric motor and gear train transmission. This problem was basically caused by friction and backlash (due to present clearance between meshed teeth), which is inherent to gear transmission. In case when gear train transmission is not present, the torque has to be controlled directly by electric motor, with high accuracy.

However, high torques are not possible without the use of electric motors and power amplifiers (which leads to an increased overall dimensions and higher mass). Moreover, electric motors and power amplifiers have unstable efficiency rates in case of short bursts of maximum electric motor power. The mentioned friction between gear teeth causes not only energy losses, but also material wear. The material wear further causes changes in component geometry which has influence on gap increase, decreasing the positioning accuracy and precision.

5.3.3 Applications of Hydraulic Control Systems

Current activity in fluid power technology includes its use to perform transmission and control functions. The growing field of robotics is giving the engineer the opportunity to perform sophisticated design studies for equipment used in many productive sectors such as aerospace, agriculture, automated manufacture, construction, defense, energy and transportation. The above gives an indication of present and future career opportunities for those with skills and experience in fluid power technology. With their increasing use, it is predicted that fluid power components will become less expensive, thereby further improving the competitive advantages of utilizing fluid power as a power transmission medium. With regard to fluid power components, considerable improvements have been made in the design of seals, fluids, valves, conductors, pumps and motors. The most significant advances in hydraulic system design, however, are seen in the area of controls. Electro-mechanical controls have diversified considerably and have led to many new hydraulic applications.

More recent developments have included the use of programmable controllers in conjunction with hydraulic systems. These controllers contain digitally operated electronic components and have programmable memory with instructions to implement functions such as logic, sequencing, timing, and counting. Such modules may control many different types of machines or processes. It is pleasing to note that fluid power applications are being extended and should increasingly improve our quality of life by, among other things, reducing the need for manual work to be performed.

Dependability has been improved by the development of easily serviced cartridge-type control valves with very long service life and minimum maintenance. Due in part to greater demand, the above systems have been reduced in cost, high pressure piping has been minimized, performance has been improved, and there has been a simplification of maintenance procedures. The improvements in physical equipment have been accompanied by an enhanced ability to analyze the performance of fluid power systems. Much of this development can be attributed to the dramatic improvement in computing power available to the engineer.

Hydraulic crane

(1) Hydraulic crane

A hydraulic crane is a type of heavy-duty equipment used for lifting and hoisting. Unlike smaller cranes, which rely on electric or diesel-powered motors, hydraulic cranes include an internal hydraulic system that allows the crane to lift heavier loads. This fluid-filled hydraulic system enables the crane to transport objects such as heavy shipping containers and tractor trailers, which are well beyond the size and scope of any other lifting device.

Each hydraulic crane can have an enclosed operator's cab set atop a steel base. As shown in Figure 5.29, this style of crane can also be mounted on top of chasis with on wheels or rollers, while other cranes are stationary. From the cab, the operator controls a large arm known as a boom. Many hydraulic cranes feature a telescoping boom, which allows the operator to reach objects from a greater distance because the boom can extend out beyond the fix length. Cables, blocks and hooks attached to the boom can be used to safely hoist or lift different equipment.

Figure 5.29 Hydraulic crane

The crane's engine powers a hydraulic pump, which applies pressure to an oil or fluid within the hydraulic system. Because oil can't be compressed, the oil transfers this applied force to other parts of the crane. By redirecting this force where it's needed to lift an object, hydraulic systems help increase power and performance.

Hydraulic cranes are rated based on their total lifting capacity, which is a factor of both their construction and the strength of the hydraulic system. A 10-ton crane for example, can lift up to 10 tons (9,070 kg). Each hydraulic crane must be chosen carefully based on the demands of a specific project, and lifting a load that's too heavy will cause the crane to fail.

Different hydraulic crane designs allow users to more easily perform specific tasks. Those on tracks or wheels may be best suited to construction sites, while many shipyards and warehouses rely on stationary cranes. Smaller hydraulic cranes can even be found on board ships or even tow trucks.

Because of the large size and power of a hydraulic crane, all operators should undergo vigorous safety training to reduce the risk of accidents. A crane that suffers operational failures could put operators or those nearby at risk from fire or falling objects. Poorly trained operators may direct the boom into nearby buildings or even people. Cranes that have not been set up properly can even tip over, leading to large-scale damage. While not all areas require safety training, individuals or organizations often pursue training to minimize liability and maximize safety.

The purpose of hydraulic cranes is to stay stable whilst lifting heavy weights. Although cranes have been used throughout the history of construction engineering, hydraulic cranes use a more technical design. The hydraulic crane depends on three separate parts when lifting incredibly heavy loads: the hydraulic cylinder, the pulley and the lever. The latter is a horizontal beam that takes on the task of the fulcrum. When a heavy object is loaded, an amount of force is applied at the other end, and in the other direction.

Known also as the jib, the pulley is a strut tilted to support a pulley block. The fixed block is wrapped with several layers of cable that is then pulled by machine or by hand. This can then create a force that is equal to the weight of the load. The hydraulic cylinder is then used to lift the load.

(2) Hydraulic lifting platform

The scissors elevator is an elevator with a system of levers and hydraulic cylinders on which the metal platform capable of moving in the vertical plane, as shown in Figure 5.30. This is achieved by using of linked folding supports in a crisscross pattern, called scissor mechanism.

Figure 5.30 Hydraulic lifting platform

The hydraulic lift is chosen as a subject because it is a perfect example of mechanical engineering field. This mechanism combines a result of several main fields of engineering and at the same time, it is simple and accessible for understanding. The construction and load distribution represent statics and strength of material subjects, the hydraulic cylinder and the control unit involve knowledge of hydraulic systems and automation. Material science is important for selection of a suitable material as well as knowledge of 3D modeling.

Also, scissors lift is an integral part of most of the workshops and building objects. The key advantage of lifts is that they even offer the best way to organize a technological and industrial process. Besides, almost all lifts give the possibility to change the place of their installation without much effort, which is important in the frequently changing conditions in the production process.

Hydraulic lifting platform

The need for the utilization of elevators is incredibly wide and it runs across workshops, factories, labs, fixing of billboards, residential/commercial buildings to repair street lights, etc. Expanded and less-efficient, the engineers may run into one or more problems while using.

The actual lifting movement is performed by the hydraulic cylinder. These are arranged inside the scissors structure so that the extending movement of the hydraulic cylinder is matched by the lifting movement of the platform. The cylinder itself is powered by means of a hydraulic pump (i.e. a gear pump). This in turn is powered by a small electric motor. In contrast to the group of mechanical lift tables, the great advantage of hydraulic lifting platforms is that the drive need not be activated for the lowering. The lowering movement is brought about solely by the weight of the lift table and possibly the weight of the load. This potential energy is sufficient to retract the hydraulic cylinder.

Hydraulic drives generally have a very high power density. This means that for relatively large force, a relatively small drive is required for a hydraulic drive. In hydraulic lifting platforms this means that the hydraulic drive for lifting the same load is more compact than, for example, a mechanical drive. This means the drive unit can often be installed in the lift table saving valuable space in production.

The concept of a scissors lift with hydraulic power comes from Pascal's law applied in car jacks and hydraulic rams which states that "pressure exerted anywhere in a conformed incompressible fluid is transmitted equally in all directions throughout the fluid such that the pressure ratio remains the same. A distinctive feature of an electro-hydraulic scissor lift in comparison with other analogues is the low price due to the use of a relatively simple design. A special lifting platform is driven by a simple metal structure with levers that look like scissors connected with others in a long chain. As a lifting force is used electro hydraulic mechanism for driving a pair of scissors in motion.

In addition, a scissor lift is suitable for use in situations, where movement of other types of lifts is limited. This capability makes this type of lift particularly versatile and convenient. A platform with load is movable not only vertically, but also on the meter to the side, as is for example available on some models. This feature is highly convenient in situations in the workplace where there is no possibility to put the basis of lift exactly under the desired object.

(3) Hydraulic machinery

Hydraulic machinery are machines and tools that use liquid fluid power to do simple work, as shown in Figure 5.31. Heavy equipment is a common example. In this type of machine, hydraulic fluid is transmitted throughout the machine to various hydraulic motors and hydraulic cylinders and becomes pressurized according to the resistance present. The fluid is controlled directly or automatically by control valves and distributed through hoses and tubes.

Figure 5.31 Hydraulic machinery

The popularity of hydraulic machinery is due to the very large amount of power that can be transferred through small tubes and flexible hoses, and the high power density and wide array of actuators that can make use of this power. Hydraulic machinery is operated by the use of hydraulics, where a liquid is the powering medium.

A fundamental feature of hydraulic systems is the ability to apply force or torque multiplication in an easy way, independent of the distance between the input and output, without the need for mechanical gears or levers, either by altering the effective areas in two connected cylinders or the effective displacement between a pump and motor. In normal cases, hydraulic ratios are combined with a mechanical force or torque ratio for optimum machine designs such as boom movements and track drives for an excavator.

For the hydraulic fluid to do work, it must flow to the actuator or motors, then return to a reservoir. The fluid is then filtered and repumped. The path taken by hydraulic fluid is called a hydraulic circuit of which there are several types. Open center circuits use pumps which supply a continuous flow. The flow is returned to tank through the control valve's open center; that is, when the control valve is centered, it provides an open return path to tank and the fluid is not pumped to a high pressure. Otherwise, if the control valve is actuated it routes fluid to and from an actuator and tank. The fluid's pressure will rise to meet any resistance, since the pump has a constant output. If the pressure rises too high, fluid returns to tank through a pressure relief valve. Multiple control valves may be stacked in series. This type of circuit can use inexpensive, constant displacement pumps. Closed center circuits supply full pressure to the control valves, whether any valves are actuated or not. The pumps vary their flow rate, pumping very little hydraulic fluid until the operator actuates a valve. The valve's spool therefore doesn't need an open center return path to tank. Multiple valves can be connected in a parallel arrangement and system pressure is equal for all valves.

(4) Applications of hydraulic systems in large engineering projects

1) Enerpac helps the Beijing's "Bird's Nest" to stand on its own feet (Figure 5.32)

In September of 2006, after two years of construction, the main venue for the 2008 Beijing Olympic Games came to a final and most important part of the construction of its steel structure: the dismantling of the temporary support towers. During construction, the crisscrossed interwoven steel roof construction of the bird's nest roof was supported by 78 temporary steel columns. For additional stability, the huge "twigs" were welded onto the supports. After completion of the bird's nest construction the "twigs" had to be cut off the support piers, before dismantling of the piers could start. In most western countries, cranes would have been hired to do the lifting job while welders would cut the welds off the 78 supports laid during construction. However, due to the extremely high cost of hiring a number of 800-ton cranes for several days in China, a smarter and less expensive solution needed to be found. Key prerequisites to the entire cutting-operation were safety, control, stability and cost. Enerpac, known from many complex hydraulic applications around the globe and in the

Chinese market especially for their hydraulic solutions for moving roofs (NanTong stadium) and moving structures (Shanghai concert hall) was consulted.

The disconnecting and dismantling process of the temporary supports came down to synchronically and fully controlled lifting of the structure off its supports, cutting the welds, followed by controlled and synchronized stage-lowering to allow the removal of the 50 mm thick leveling plates that were used during construction. Computer-controlled hydraulics is the perfect match for jobs like this, and Enerpac was granted the contract to perform the stage lifting and lowering of the roof.

The entire configuration including the central computer, satellite computer-controllers, 156 double-acting high-pressure hydraulic cylinders and 55 electronically controlled hydraulic power units was specified and custom designed by Enerpac. For added safety, control, and accuracy, multi-functional valves, load sensors, stroke sensors, shift detection and a digital feedback system were integrated.

The design of the bird's nest is based on three construction circles: an outer circle, a central circle and an inner circle. Each circle has a specific number of supporting piers, varying from 24 for the outer and central circles to 30 for the inner circle. For load, control and accuracy reasons the 78 support points including their hydraulic systems are divided into 10 regions, each of which has its own satellite controller. For the actual stage lifting and lowering process each support pier is equipped with two 150-ton double-acting cylinders. At the central computer all load and stroke data are pre-programmed for a fully controlled lifting and lowering process. During the stage lowering process the bird's nest is alternatively supported by the hydraulic cylinders and the leveling plates on the temporary supports.

After successfully disconnecting the 45,000 ton steel structure from its temporary support piers, the bird's nest stood on its own "feet" for the first time.

2) Hydraulic jacking system assists construction of the Honolulu rail transit system (shown in Figure 5.32)

In 2012, the Honolulu rail transit project began the construction of a 20-mile elevated rail line. The rail system seeks to alleviate significant traffic issues affecting Hawaii's most congested city. Kiewit was tasked with the design and construction of the elevated rail guideway. In order to accelerate the construction timeline and maintain a high level of quality, Kiewit selected a segmental precast span-by-span construction technique to complete the project. The first 10 miles of guideway will consist of 5,200 guideway precast

Figure 5.32　Hydraulic jacking system

segments which are cast at an off-site casting yard, then delivered onsite for installation. Kiewit designed a twin underslung girder system to complete the span-by-span construction. The installation of segments weighing up to 50 tons each requires a positioning system with a significant amount of force and precision.

Enerpac engineered and delivered the hydraulic control package to Kiewit for the underslung girder systems. Each of the twin underslung girder systems are supported by pier brackets on each end which are transversely positioned and aligned with custom Enerpac TR-series hydraulic tie rod cylinders. Once the steel box girders are launched and aligned on the pier brackets, custom Enerpac 150-ton locknut cylinders are mechanically locked ensuring the system maintains its precise position. The guideway segments are then picked via crane and placed onto a set of segment carts where they are supported by 50-ton custom CLL-series hydraulic locknut cylinders. Utilizing the segment carts the segments are precisely aligned and maneuvered into position, where the segments can then be epoxied and post-tensioned together completing the span. Span-by-span advancement is enabling multiple phases of the project to occur simultaneously, thus helping to reduce the overall project timeline.

Words and Expressions

bulk modulus [bʌlk ˈmɒdjʊləs]	体积模量
bandwidth [ˈbændwɪdθ] a data transmission rate; the maximum amount of information (bits/second) that can be transmitted along a channel	n. 带宽
relief valve [rɪˈliːf vælv]	安全阀
Reynolds number [ˈrenəldz ˈnʌmbə]	雷诺数
gear pump [ɡɪə pʌmp]	齿轮泵
bearing [ˈbeərɪŋ] relevant relation or interconnection	轴承
pump controlled system [pʌmp kənˈtrəʊld ˈsɪstəm]	泵控系统
valve controlled system [vælv kənˈtrəʊld ˈsɪstəm]	阀控系统
vane pump [veɪn pʌmp]	叶轮泵
axial piston pump [ˈæksiːəl ˈpɪstən pʌmp]	轴向柱塞泵
swashplate [ˈswɒʃpleɪt] the rotating shaft and plate are shown in silver	n. 斜盘
radial piston pump [ˈreɪdiːəl ˈpɪstən pʌmp]	径向柱塞泵

spool [spu:l] a winder around which thread or tape or film or other flexible materials can be wound	n. 阀芯
hydraulic cylinder [haɪˈdrɔːlɪk ˈsɪlɪndə]	液压缸
reservoir [ˈrezəvwɑː(r)] a large or extra supply of something	储油器
accumulator [əˈkjuːmjəleɪtə(r)] a voltaic battery that stores electric charge	蓄能器
filter [ˈfɪltə] a device that removes something from whatever passes through it	过滤器
O-ring seal [ˈəʊˌrɪŋ ˈsiːl]	O 形圈密封
weld [ˈweld] fastening two pieces of metal together by softening with heat and applying pressure	v. 焊接
pipe fitting [paɪp ˈfɪtɪŋ]	管配件
electro-hydrostatic actuators (EHA) [ɪˈlektrəʊ ˈhaɪdrəʊˈstætɪk ˈæktʃʊeɪtə]	电动静液作动器
fly-by-wire (FBW) [flaɪ baɪ ˈwaɪə(r)]	电传飞行
power-by-wire (PBW) [ˈpaʊə baɪ ˈwaɪə]	功率电传
actuator [ˈæktjuːˌeɪtə] a mechanism that puts something into automatic action	n. 作动器
flight control system [flaɪt kənˈtrəʊl ˈsɪstəm]	飞行控制系统
servo motor [ˈsɜːvəʊ ˈməʊtə]	伺服马达
reversing pump [rɪˈvɜːsɪŋ pʌmp]	反向泵
double-acting hydraulic cylinder [ˈdʌbl ˈæktɪŋ haɪˈdrɔːlɪk ˈsɪlɪndə]	双作用液压缸
anodized housing [ˈænədaɪzd ˈhaʊzɪŋ]	阳极氧化外壳
stainless steel rod [ˈsteɪnlɪs stiːl rɒd]	不锈钢杆
power amplifier [ˈpaʊə ˈæmplɪfaɪə]	功率放大器
hydraulic crane [haɪˈdrɔːlɪk kreɪn]	液压吊车
tractor trailer [ˈtræktə ˈtreɪlə(r)]	牵引拖车

jib [dʒɪb] any triangular fore-and-aft sail (set forward of the foremast)	n. 悬臂
pulley [ˈpʊli] a simple machine consisting of a wheel with a groove in which a rope can run to change the direction or point of application of a force applied to the rope	n. 滑轮
hydraulic lifting platform [haɪˈdrɔːlɪk ˈlɪftɪŋ ˈplætfɔːm]	液压升降平台
scissors elevator [ˈsɪzəz ˈelɪveɪtə]	剪刀式升降机
billboard [ˈbɪlbɔːd] large outdoor signboard	n. 广告牌
car jack [kɑːˈdʒæk]	汽车千斤顶
hydraulic ram [haɪˈdrɔːlɪk ræm]	液压油缸
excavator [ˈekskəveɪtə] a machine for excavating	n. 挖掘机
multifunctional valve [ˌmʌltiˈfʌŋkʃənl vælv]	多功能阀
hydraulic jacking system [haɪˈdrɔːlɪk ˈdʒækɪŋ ˈsɪstəm]	液压顶升系统
bracket [ˈbrækɪt] a category falling within certain defined limits	n. 臂架

Notes

[1] 液压传动是指以液体为工作介质进行能量传递和控制的一种传动方式。在液体传动中，根据其能量传递形式不同，又分为液力传动和液压传动。液力传动主要是利用液体动能进行能量转换的传动方式。

[2] 液压控制系统是以电机提供动力基础，使用液压泵将机械能转化为压力，推动液压油，通过控制各种阀门改变液压油的流向，从而推动液压缸做出不同行程、不同方向的动作，完成各种设备不同的动作需要。

[3] 电动静液作动器，是功率电传的典型代表，一般采用一体化集成设计。由于采用功率电传而不是集中液压控制系统，即连接各个液压作动器系统的只有电缆，不再需要集中的液压能源站，减少了液压管路与液压设备，降低了因油液的泄漏与电磁阀控制所造成的能耗损失，同时具有传统液压系统功率大的优点，成为机电液一体化研究的热点。

[4] 伺服电机可使控制速度、位置精度非常准确，可以将电压信号转化为转矩和转速以驱动控制对象。伺服电机转子转速受输入信号控制，并能快速反应，在自动控制系统中，用作执行元件，且具有机电时间常数小、线性度高、始动电压等特性，可把所收到的电信号转换成电动机轴上的角位移或角速度输出。

Questions for Discussion

1. What are the advantages and disadvantages of the hydraulic control systems?
2. What is the difference between the pump controlled system and the valve controlled system?
3. List several types of hydraulic pumps.
4. What parts does the EHA system diagram contain?
5. What are the configurations of the EHA system?
6. How to apply the EHA system to the robot?
7. What is the purpose of hydraulic cranes?
8. List some other applications of hydraulic systems.
9. Which three structures circles are the design of the bird's nest based on?

5.4 Embedded System and Intelligent Instrument

5.4.1 Embedded System

(1) The definition of embedded system

An embedded system is a programmed controlling and operating system with a dedicated function within a larger mechanical or electrical system, often with real-time computing constraints. It is embedded as part of a complete device often including hardware and mechanical parts. Embedded systems control many devices in common use today. Ninety-eight percent of all microprocessors are manufactured as components of embedded systems.

Modern embedded systems are often based on microcontrollers (i.e. CPUs with integrated memory or peripheral interfaces, as shown in Figure 5.33), but ordinary microprocessors (using external chips for memory and peripheral interface circuits) are also common, especially in more-complex systems. In either case, the processor(s) used may be types ranging from general purpose to those specialized in certain class of computations, or even custom designed for the application at hand. A common standard class of dedicated processors is the digital signal processor (DSP).

(2) The applications of embedded system

Embedded systems range from portable devices such as digital watches, to large stationary installations like traffic lights, factory controllers, and largely complex systems like hybrid vehicles, MRI, and avionics. Complexity varies from low, with a single microcontroller chip, to very high with multiple units, peripherals and networks mounted inside a large chassis or enclosure.

Figure 5.33 An embedded system on a plug-in card with processor, memory, power supply, and peripheral interfaces

Embedded systems are commonly found in consumer, cooking, industrial, automotive, medical, commercial and military applications. Transportation systems from flight to automobiles increasingly use embedded systems. New airplanes contain advanced avionics such as inertial guidance systems and GPS receivers that also have considerable safety requirements. Various electric motors (brushless DC motors, induction motors and DC motors) use electric motor controllers. Automobiles, electric vehicles, and hybrid vehicles increasingly use embedded systems to maximize efficiency and reduce pollution. Other automotive safety systems include anti-lock braking system (ABS), electronic stability control (ESC), traction control (TCS) and automatic four-wheel drive.

(3) The characteristics of embedded system

Embedded systems are designed to do some specific task, rather than be a general-purpose computer for multiple tasks. Some also have real-time performance constraints that must be met, for reasons such as safety and usability; others may have low or no performance requirements, allowing the system hardware to be simplified to reduce costs.

1) User interface

Embedded systems range from no user interface at all, in systems dedicated only to one task, to complex graphical user interfaces that resemble modern computer desktop operating systems.

2) Processors in embedded systems

Embedded processors can be broken into two broad categories. Ordinary microprocessors (μP) use separate integrated circuits for memory and peripherals. Microcontrollers (μC) have on-chip peripherals, thus reducing power consumption, size and cost.

(4) Ready-made computer boards

PC/104 and PC/104+ are examples of standards for ready-made computer boards intended for small, low-volume embedded and ruggedized systems, mostly x86-based. These are often physically small compared to a standard PC, although still quite large compared to most simple (8/16-bit) embedded systems. They often use DOS, Linux, NetBSD, or an embedded real-time operating system such as MicroC/OS-II, QNX or VxWorks. Sometimes these boards use non-x86 processors.

(5) ASIC and FPGA solutions

A common array for very-high-volume embedded systems is the system on a chip (SoC) that contains a complete system consisting of multiple processors, multipliers, caches and interfaces on a single chip. SoCs can be implemented as an application-specific integrated circuit (ASIC) or using a field-programmable gate array (FPGA).

(6) Peripherals

Embedded systems talk with the outside world via peripherals, such as:
- Serial communication interfaces (SCI): RS-232, RS-422, RS-485, etc.;
- Synchronous serial communication interface: I2C, SPI, SSC and ESSI (Enhanced Synchronous Serial Interface);
- Universal serial bus (USB);
- Multi media cards (SD cards, compact flash, etc.);
- Networks: Ethernet, LonWorks, etc.;
- Fieldbuses: CAN-bus, LIN-bus, profibus, etc.;
- Timers: PLL(s), capture/compare and time processing units;
- Discrete IO: general purpose input/output (GPIO);
- Analog to digital converter/digital to analog converter (ADC/DAC);
- Debugging: JTAG, ISP, ICSP, BDM port, BITP, and DB9 ports.

(7) Tools

As with other software, embedded system designers use compilers, assemblers, and debuggers to develop embedded system software. However, they may also use some more specific tools as follows:
- In circuit debuggers or emulators (see next section);
- Utilities to add a checksum or CRC to a program, so the embedded system can check if the program is valid;
- For systems using digital signal processing, developers may use a math workbench to simulate the mathematics.

Software tools can come from several sources as follows:
- Software companies that specialize in the embedded market;
- Ported from the GNU software development tools;
- Sometimes, development tools for a personal computer can be used if the embedded

processor is a close relative to a common PC processor.

(8) Debugging

Embedded debugging may be performed at different levels, depending on the facilities available. The different metrics that characterize the different forms of embedded debugging are: does it slow down the main application, how close is the debugged system or application to the actual system or application, how expressive are the triggers that can be set for debugging (i.e. inspecting the memory when a particular program counter value is reached), and what can be inspected in the debugging process (i.e. only memory, or memory and registers, etc.).

1) Tracing

Real-time operating systems (RTOS) often supports tracing of operating system events. A graphical view is presented by a host PC tool, based on a recording of the system behavior. The trace recording can be performed in software, by the RTOS, or by special tracing hardware. RTOS tracing allows developers to understand timing and performance issues of the software system and gives a good understanding of the high-level system behaviors. Commercial tools like RTXC quadros or IAR systems exists.

2) Reliability

Embedded systems often reside in machines that are expected to run continuously for years without errors, and in some cases recover by themselves if an error occurs. Therefore, the software is usually developed and tested more carefully than that for personal computers, and unreliable mechanical moving parts such as disk drives, switches or buttons are avoided.

3) High vs. low volume

For high volume systems such as portable music players or mobile phones, minimizing cost is usually the primary design consideration. Engineers typically select hardware that is just "good enough" to implement the necessary functions.

For low-volume or prototype embedded systems, general purpose computers may be adapted by limiting the programs or by replacing the operating system with a real-time operating system.

(9) Embedded software architectures

There are several different types of software architecture in common use.

1) Simple control loop

In this design, the software simply has a loop. The loop calls subroutines, each of which manages a part of the hardware or software. Hence it is called a simple control loop or control loop.

2) Interrupt-controlled system

Some embedded systems are predominantly controlled by interrupts. This means that tasks performed by the system are triggered by different kinds of events; an interrupt could be generated, for example, by a timer in a predefined frequency, or by a serial port controller receiving a byte.

3) Cooperative multitasking

A nonpreemptive multitasking system is very similar to the simple control loop scheme, except that the loop is hidden in an API. The programmer defines a series of tasks, and each task gets its own environment to "run" in. When a task is idle, it calls an idle routine, usually called "pause", "wait", "yield", "nop" (stands for no operation), etc.

4) Preemptive multitasking or multi-threading

In this type of system, a low-level piece of code switches between tasks or threads based on a timer (connected to an interrupt). This is the level at which the system is generally considered to have an "operating system" kernel. Depending on how much functionality is required, it introduces more or less of the complexities of managing multiple tasks running conceptually in parallel.

5) Microkernels and exokernels

A microkernel is a logical step up from a real-time OS. The usual arrangement is that the operating system kernel allocates memory and switches the CPU to different threads of execution. User mode processes implement major functions such as file systems, network interfaces, etc.

6) Monolithic kernels

In this case, a relatively large kernel with sophisticated capabilities is adapted to suit an embedded environment. This gives programmers an environment similar to a desktop operating system like Linux or Microsoft Windows, and is therefore very productive for development; on the downside, it requires considerably more hardware resources, is often more expensive, and, because of the complexity of these kernels, can be less predictable and reliable.

7) Additional software components

In addition to the core operating system, many embedded systems have additional upper-layer software components. These components consist of networking protocol stacks like CAN, TCP/IP, FTP, HTTP, and HTTPS, and also included storage capabilities like FAT and flash memory management systems. If the embedded device has audio and video capabilities, then the appropriate drivers and codecs will be present in the system. In the case of the monolithic kernels, many of these software layers are included. In the RTOS category, the availability of the additional software components depends upon the commercial offering.

5.4.2 Intelligent Instrumentation

What is intelligent instrumentation? An instrumentation system is a physical system which is a collection of physical objects connected in such a way to get the desired output for a given input condition. Such instrumentation is termed as intelligent instrumentation when the system is made capable of automating the process by just simply feeding in a software program in it and secondly capable of storing the processed data in its memory for its further use. Thus a system is said to be an intelligent instrumentation system when it is included with a microprocessor and memory thus acting as a microcomputer. Actually, embedded system is an essential component of intelligent instrumentation. Noting that:

- The term intelligent when applied to measurement, or control, systems, means that a microprocessor or computer is used for signal processing.
- The term dumb (non-intelligent) is applied to conventional measurement systems when no such microprocessor is used. Signal processing with a microprocessor requires the signals to be digital rather than analogue.
- With a digital signal information is transmitted in the form of pulses, i.e. on-off or high-low signals.
- The off or low signal is represented by the binary number 0 and the on or high signal by the binary number 1 each such number is called a bit.
- A set of such binary no's is called a word.
- Many transducers are analogue devices. i.e. their output is some replica or scaling of the input.
- An obvious example of an analogue device is a watch where the time is represented by the position of the watch hands, while the digital watch represents time by a sequence of numbers.
- Because of this analogue nature of many transducers, intelligent instrumentation require a signal conditioning element which converts the analogue transducer output to digital form before it can be fed to the microprocessor[1].

(1) Intelligent instruments

The impact that the microprocessor has had on the domestic consumer is very evident, with its inclusion in washing machines, vehicle fuel control systems and home computers to name but a few applications. The advent of the microprocessor has had an equally significant, though perhaps not so obviously apparent, impact on the field of instrumentation[2]. It is therefore fitting that a section should be devoted to describing the principles of operation of microprocessors and explaining how they are included and programmed as a microcomputer within a measuring instrument, thereby producing what has come to be known as an intelligent instrument.

An intelligent instrument comprises all the usual elements of a measurement system and is only distinguished from dumb measurement systems by the inclusion of a microprocessor to fulfill the signal processing function (shown in Figure 5.34). The effect of this computerization of the signal processing function is an improvement in the quality of the instrument output measurements and a general simplification of the signal processing task. Some examples of the signal processing which a microprocessor can readily perform include correction of the instrument output for bias caused by environmental variations (i.e. temperature changes), and conversion to produce a linear output from a transducer whose characteristic is fundamentally non-linear.

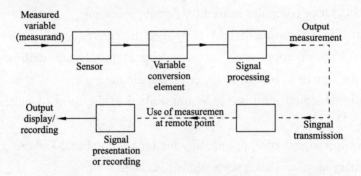

Figure 5.34 Signal processing in intelligent instruments

(2) Elements of a microcomputer

The primary function of a digital computer is the manipulation of data. The three elements which are essential to the fulfillment of this task are the central processing unit, the memory and the input-output interface. These elements are collectively known as the computer hardware, and each element exists physically as one or more integrated circuit chips mounted on a printed circuit board.

(3) Number systems

As we all know, the fundamental role of a computer is the manipulation of data. Numbers are used both in quantifying items of data and in the form of codes which define the computational operations to be executed. All numbers which are used for these two purposes must be stored within the computer memory and also transported along the communication buses. A detailed consideration of the conventions used for representing numbers within the computer is therefore required.

(4) Programming and program execution

In most modes of usage, including use as part of intelligent instruments, computers are involved in manipulating data. This requires data values to be input, processed and output according to a sequence of operations defined by the computer program.

(5) Computer interfacing

The input-output interface connects the computer to the outside world, and is therefore an essential part of the computer system. When the CPU puts the address of a peripheral on to the address bus, the input-output interface decodes the address and identifies the unique computer peripheral with which a data transfer operation is to be executed. The interface also has to interpret the command on the control bus so that the timing of the data transfer is correct. One further very important function of the input-output interface is to provide a physical electronic highway for the flow of data between the computer data bus and the external peripheral.

(6) Computer address decoding

A typical address bus in a microcomputer is 16 bits wide, allowing 65,536 separate addresses to be accessed in the range 0000~FFFF (in hexadecimal representation). Special commands on some computers are reserved for accessing the bottom-end 256 of these addresses in the range 0000~00FF, and, if these commands are used, only 8 bits are needed to specify the required address. For the purpose of explaining address-decoding techniques, the scheme below shows how the lower 8 bits of the 16 bit address line are decoded to identify the unique address referenced by one of these special commands. Decoding of all 16 address lines follows a similar procedure but requires a substantially greater number of integrated circuit chips.

(7) Data transfer control

The transfer of data between the computer and peripherals is managed by control and status signals carried on the control bus which determine the exact sequencing and timing of I/O operations. Such management is necessary because of the different operating speeds of the computer and its peripherals and because of the multi-tasking operation of many computers. This means that, at any particular instant when a data transfer operation is requested, either the computer or the peripheral may not be ready to take part in the transfer.

(8) Analog to digital conversion (A/D)

Many computer inputs, particularly those from transducers within intelligent instruments, consist of analog signals which must be converted to a digital form before they can be accepted by the computer. This conversion is performed by an analog to digital conversion circuit within the computer interface.

(9) Digital to analog conversion (D/A)

Digital to analog conversion is much simpler to achieve than analog to digital conversion and the cost of building the necessary hardware circuit is considerably less. It is required wherever the output of an intelligent instrument needs to be presented on a display device which operates in an analog manner.

(10) Intelligent instruments in use

The intelligent instrument behaves as a black box as far as the user is concerned, and no knowledge of its internal mode of operation is required in normal measurement situations.

Intelligent instruments offer many advantages over their non-intelligent counterparts, principally because of the improvement in accuracy achieved by processing the output of transducers to correct for errors inherent in the measurement process[3]. Proper procedures must always be followed in their use to avoid the possibility of introducing extra sources of measurement error.

Words and Expressions

embedded system [ɪmˈbedɪd ˈsɪstəm]	嵌入式系统
intelligent instrument [ɪnˈtelɪdʒənt ˈɪnstrʊmənt]	智能仪表
dedicated function [ˈdedɪkeɪtɪd ˈfʌŋkʃən]	专用功能
real-time [ˈrɪəl ˈtaɪm]	实时
microprocessor [ˈmaɪkrəʊˌprəʊsesə] integrated circuit semiconductor chip that performs the bulk of the processing and controls the parts of a system	n. 微处理器
microcontroller [ˈmaɪkrəʊkənˈtrəʊlə] micro controllers	n. 微控制器
peripheral interface [pəˈrɪfərəl ˈɪntəfeɪ]	外设接口
digital signal processor (DSP) [ˈdɪdʒɪtl ˈsɪgnəl ˈprəʊsesə]	数字信号处理器
plug-in card [ˈplʌgˌɪn kɑːd]	插件卡
portable [ˈpɔːtəbl] a small light typewriter; usually with a case in which it can be carried	adj. 轻便的
stationary [ˈsteɪʃənəri] standing still	adj. 固定的
MRI magnetic resonance imaging	磁共振成像
hybrid vehicle [ˈhaɪbrɪd ˈviːɪkl]	混合动力汽车
avionics [ˌeɪviˈɒnɪks] science and technology of electronic systems and devices for aeronautics and astronautics	n. 航空电子设备

chassis or enclosure [ˈʃæsi ɔː ɪnˈkləʊʒə]	底盘或圈地
inertial guidance system and GPS receiver [ɪˈnɜːʃl ˈɡaɪdəns sɪstəm ænd dʒiː piː es rɪˈsiːvə]	惯导系统和GPS接收机
brushless DC motor [brʌʃlɪs diː siː məʊtə]	无刷直流电机
induction motor [ɪnˈdʌkʃən məʊtə]	感应电机
anti-lock braking system (ABS) [ˈæntɪˌlɒk ˈbreɪkɪŋ ˈsɪstəm]	防抱死制动装置
traction control system (TCS) [ˈtrækʃən kənˈtrəʊl ˈsɪstəm]	牵引控制系统
user interface [ˈjuːzə ˈɪntəfeɪs]	用户接口
ready-made [ˌrediˈmeɪd]	*adj.* 现成的
low-volume embedded and ruggedized system [ləʊ ˈvɒljuːm ɪmˈbedɪd ænd ˈrʌɡɪdaɪzd ˈsɪstəm]	低容量嵌入式加固系统
multiplier [ˈmʌltɪplaɪə(r)] the number by which a multiplicand is multiplied	*n.* 乘法器
cache [kæʃ] a hidden storage space	*n.* 高速缓存
application-specific integrated circuit (ASIC) [ˌæplɪˈkeɪʃən spɪˈsɪfɪk ˈɪntɪɡreɪtɪd ˈsɜːkɪt]	专用集成电路
field-programmable gate array (FPGA) [fiːld ˈprəʊɡræməbl ɡeɪt əˈreɪ]	现场可编程门阵列
compiler [kəmˈpaɪlə] a program that decodes instructions written in a higher order language and produces an assembly language program	*n.* 编译器
assembler [əˈsemblə] a program to convert assembly language into machine language	*n.* 汇编器
debugger [diːˈbʌɡə] a program that helps in locating and correcting programming errors	*n.* 调试器
emulator [ˈemjuleɪtə(r)] someone who copies the words or behavior of another	*n.* 仿真器

trigger [ˈtrɪgə] a device that activates or releases or causes something to happen	n. 触发器
register [ˈredʒɪstə(r)] an official written record of names or events or transactions	n. 寄存器
checksum [ˈtʃeksʌm] a digit representing the sum of the digits in an instance of digital data; used to check whether errors have occurred in transmission or storage CRC	n. 校验和；循环冗余码校验
workbench [ˈwɜːkbentʃ] a strong worktable for a carpenter or mechanic	n. 工作台
real-time operating system (RTOS) [rɪəl taɪm ˈɒpəˌreɪtɪŋ ˈsɪstəm]	实时操作系统
reside in [rɪˈzaɪd ɪn]	嵌入
software architecture [ˈsɒftweə ˈɑːkɪtektʃə]	软件架构
subroutine [ˈsʌbruːtiːn] a set sequence of steps, part of larger computer program	n. 子程序
interrupt-controlled [ˌɪntəˈrʌpt kənˈtrəʊld]	中断控制
predominantly [prɪˈdɒmɪnəntli] much greater in number or influence	adv. 主要地
predefined [ˌpriːdɪˈfaɪnd] define sth. in advance	adj. 预定义
cooperative multitasking [kəʊˈɒpərətɪv ˌmʌltiˈtɑːskɪŋ]	协同多任务
nonpreemptive [nɒn priːˈemptɪv]	不可中断
API (application program interface) [ˌæplɪˈkeɪʃən ˈprəʊgræm ˈɪntəfeɪs]	应用程序接口
idle [ˈaɪdl] the state of an engine or other mechanism that is idling	n. 空闲
preemptive multitasking or multi-threading [priːˈemptɪv ˌmʌltiˈtɑːskɪŋ ɔː ˈmʌlti ˈθredɪŋ]	抢占式多任务或多线程
kernel [ˈkɜːnl] the inner and usually edible part of a seed or grain or nut or fruit stone	n. 核

conceptually [kənˈseptʃuəli] in a conceptual manner	*adv.* 概念上
microkernel and exokernel [maɪkˈrəʊkɜːnl ænd eksˈtɜːnl]	微内核与外层内核
monolithic kernel [ˌmɒnəˈlɪθɪk ˈkɜːnəl]	整体内核
sophisticated [səˈfɪstɪkeɪtɪd] having or appealing to those having worldly knowledge and refinement and savoir-faire	*adj.* 复杂巧妙的；先进的
considerably [kənˈsɪdərəbli] to a great extent or degree	*adv.* 相当地
networking protocol stack [ˈnetwɜːkɪŋ ˈprəʊtəkɒl stæk]	网络协议栈
codec [kəʊdeks] a device that changes an electronic signal into a form that people can understand	*n.* 解码器
dumb [dʌm] lacking intellectual acuity	*adj.* 非智能的
binary number [ˈbaɪnəri ˈnʌmbə]	二进制数
transducer [trænzˈdjuːsə(r)] an electrical device that converts one form of energy into another	*n.* 传感器；换能器
replica [ˈreplɪkə] copy that is not the original; something that has been copied	*n.* 复制品
convert [kənˈvɜːt] change the nature, purpose, or function of something	*v.* 转变；转换
computerization [kəmˌpjuːtəraɪˈzeɪʃən] the control of processes by computer	*n.* 计算机化
readily [ˈredɪli] without much difficulty	*adv.* 轻而易举地
fundamentally [fʌndəˈmentəli] at bottom or by one's (or its) very nature	*adv.* 根本上
printed circuit board (PCB) [ˈprɪntɪd ˈsɜːkɪt bɔːd]	印制电路板
address bus [əˈdres bʌs]	地址总线
data bus [ˈdeɪtə bʌs]	数据总线

control bus [kənˈtrəʊl bʌs]	控制总线
decode [ˌdiːˈkəʊd] convert code into ordinary language	v. 解码
binary [ˈbaɪnəri] a system of two stars that revolve around each other under their mutual gravitation	n. 二进制
decimal [ˈdesɪml] a proper fraction whose denominator is a power of 10	n. 十进制
analog to digital conversion [ˈænəlɒg tuː ˈdɪdʒɪtl kənˈvɜːʃən]	A/D 模数转换
digital to analog conversion [ˈdɪdʒɪtl tuː ˈænəlɔːg kənˈvɜːʃən]	D/A 数模转换

Notes

[1] 本句可译为：由于多数传感器具有这种模拟特性，智能仪表需要一个信号调节元件，将模拟的传感器输出转换为数字形式，然后才能输入到微处理器。

[2] 本句可译为：微处理器的出现在仪器仪表领域产生了同样重要的影响，尽管影响可能没那么明显。

[3] 本句可译为：与非智能仪器相比，智能仪器具有许多优势，主要是因为通过处理传感器的输出来校正测量过程中固有的误差，从而提高了精度。

Questions for Discussion

1. What's the relationship between embedded system and intelligent instrument?

2. How many components constitute an embedded system? How about intelligent instrument?

3. Can you draw out a diagram of embedded system and intelligent instrument?

4. How many kinds of number representation do you know? Can you write out the conversion formula among binary, decimal, and hex number?

Chapter 6

Vehicle Engineering

6.1 Vehicle Structure Reliability

Until the 1960s, quality targets were deemed to have been reached when the item considered was found to be free of defects or systematic failures at the time it left the manufacturer. The growing complexity of equipment and systems, as well as the rapidly increasing cost incurred by loss of operation as a consequence of failures, have brought to the forefront the aspects of reliability, maintainability, availability, and safety. The expectation today is that complex equipment and systems are not only free from defects and systematic failures at time $t = 0$ (when they are put into operation), but also perform the required function failure free for a stated time interval and have a fail-safe behavior in case of critical or catastrophic failures[1]. However, the question of whether a given item will operate without failures during a stated period of time cannot be simply answered by yes or no, on the basis of a compliance test. Experience shows that only a probability for this occurrence can be given. This probability is a measure of the item's reliability and can be interpreted as follows:

If n statistically identical and independent items are put into operation at time $t = 0$ to perform a given mission and $\bar{v} \leq n$ of them accomplish it successfully, then the ratio $\bar{v}/n$ is a random variable which converges for increasing n to the true value of the reliability.

Performance parameters as well as reliability, maintainability, availability, and safety have to be built in during design and development and retained during production and operation of the item.

6.1.1 Basic Concepts

(1) Reliability

Reliability is a characteristic of the item, expressed by the probability that it will perform its required function under given conditions for a stated time interval. It is generally designated by R. From a qualitative point of view, reliability can be defined as the ability of the item to remain functional. Quantitatively, reliability specifies the probability that no operational interruptions will occur during a stated time interval. This does not mean that redundant parts may not fail, such parts can fail and be repaired (without operational interruption at item/system level). The concept of reliability thus applies to nonrepairable as well as to repairable items. To make sense, a numerical statement of reliability (i.e. $R = 0.9$) must be accompanied by the definition of the required function, the operating conditions, and the mission duration. In general, it is also important to know whether or not the item can be considered new when the mission starts.

An item is a functional or structural unit of arbitrary complexity (i.e. component, assembly, equipment, subsystem, system) that can be considered as an entity for investigations. The required function or operating conditions can be time dependent. In these cases, a mission profile has to be defined and all reliability figures will be related to it. A representative mission profile and the corresponding reliability targets should be given in the item's specifications.

Often the mission duration is considered as a parameter t, the reliability function is then defined by $R(t)$. $R(t)$ is the probability that no failure at item level will occur in the interval $(0, t)$.

(2) Failure

A failure occurs when the item stops performing its required function. As simple as this definition is, it can become difficult to apply it to complex items. The failure free time is generally a random variable. It is often reasonably long; but it can be very short, for instance because of a failure caused by a transient event at turn-on. A general assumption in investigating failure-free times is that at $t = 0$ the item is free of defects and systematic failures. Besides their frequency, failures should be classified according to the mode, cause, effect, and mechanism.

- Mode: The mode of a failure is the symptom (local effect) by which a failure is observed (i.e. brittle rupture, creep, cracking, seizure, fatigue for mechanical components).
- Cause: The cause of a failure can be intrinsic, due to weaknesses in the item or wear out, or extrinsic, due to errors, misuse or mishandling during the design, production, or use.

Extrinsic causes often lead to systematic failures, which are deterministic and should be considered like defects.
- Effect: It can be different if considered on the item itself or at higher level. A usual classification is non relevant, partial, complete, and critical failure. Since a failure can also cause further failures, distinction between primary and secondary failure is important.
- Mechanism: Failure mechanism is the physical, chemical, or other process resulting in a failure.

(3) Failure rate, MTTF, MTBF

The failure rate $\lambda(t)$ of a large population of statistically identical and independent items exhibits often a typical bathtub curve (shown in Figure 6.1) with the following 3 phases:
- Early failures: $\lambda(t)$ decreases (in general) rapidly with time; failures in this phase are attributable to randomly distributed weaknesses in materials, components, or production processes.
- Failures with constant (or nearly so) failure rate: $\lambda(t)$ is approximately constant; failures in this period are Poisson distributed and often cataleptic.
- Wear out failures: $\lambda(t)$ increases with time; failures in this period are attributable to aging, wear out, fatigue, etc. (i. e. corrosion, electro migration).

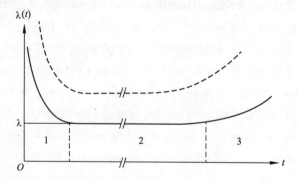

Figure 6.1 Typical shape for the failure rate of a large population of statistically identical and independent (no repairable) items (dashed is a possible shift for a higher stress, e.g., ambient temperature)

Mean time to failure (MTTF) is a basic measure of reliability for non-repairable systems. It is the mean time expected until the first failure of a piece of equipment. MTTF is a statistical value and is meant to be the mean over a long period of time and a large number of units.

Mean time between failure (MTBF) is a reliability term used to provide the amount of failures per million hours for a product. This is the most common inquiry about a product's life span, and is important in the decision-making process of the end user. MTBF is more important for industries and integrators than for consumers. Most consumers are price driven and will not take MTBF into consideration, nor is the data often readily available. On the other hand, when equipment such as media converters or switches must be installed into mission critical applications, MTBF becomes very important.

The MTBF is often calculated based on an algorithm that factors in all of a product's components to reach the sum life cycle in hours. In reality, depreciation modes of the product could limit the life of the product much earlier. It is very possible to have a product with an extremely high MTBF, but an average or more realistic expected service life.

Technically, MTBF should be used only in reference to a repairable item, while MTTF should be used for non-repairable items. However, MTBF is commonly used for both repairable and non-repairable items.

(4) Maintenance, maintainability

Maintenance defines the set of actions performed on the item to retain it in or to restore it to a specified state. Maintenance is thus subdivided into preventive maintenance, carried out at predetermined intervals to reduce wear out failures, and corrective maintenance, carried out after failure detection and intended to put the item into a state in which it can again perform the required function[2]. Aim of a preventive maintenance is also to detect and repair hidden failures, i.e., failures in redundant elements not detected at their occurrence. Corrective maintenance is also known as repair, and can include any or all of the following steps: detection, localization (isolation), correction, and checkout. To simplify calculations, it is generally assumed that the element in the reliability block diagram for which a maintenance action has been performed is as-good-as-new after maintenance. This assumption is valid for the whole equipment or system in the case of constant failure rate for all elements which have not been repaired or replaced.

Maintainability is a characteristic of the item, expressed by the probability that a preventive maintenance or a repair of the item will be performed within a stated time interval for given procedures and resources (skill level of personnel, spare parts, test facilities, etc.). From a qualitative point of view, maintainability can be defined as the ability of the item to be retained in or restored to a specified state. Maintainability has to be built into complex equipment and systems during design and development by realizing a maintenance concept. Due to the increasing maintenance cost, maintainability aspects have grown in importance.

6.1.2 Historical development

Methods and procedures of quality assurance and reliability engineering have been developed extensively. For indicative purpose, Table 6.1 summarizes major steps of this development and Figure 6.2 shows the approximate distribution of the effort between quality assurance and reliability engineering during the same period of time.

Table 6.1 Historical development of quality assurance and reliability engineering

Time	Historical development
Before 1940	Quality attributes and characteristics are defined. In-process and final tests are carried out, usually in a department within the production area. The concept of quality of manufacture is introduced
1940–1950	Defects and failures are systematically collected and analyzed. Corrective actions are carried out. Statistical quality control is developed. It is recognized that quality must be built into an item. The concept quality of design becomes important
1950–1960	Quality assurance is recognized as a means for developing and manufacturing an item with a specified quality level. Preventive measures (actions) are added to tests and corrective actions. It is recognized that correct short-term functioning does not also signify reliability. Design reviews and systematic analysis of failures (failure data and failure mechanisms), performed often in the research and development area, lead to important reliability improvements
1960–1970	Difficulties with respect to reproducibility and change control, as well as interfacing problems during the integration phase, require a refinement of the concept of configuration management. Reliability engineering is recognized as a means of developing and manufacturing an item with specified reliability. Reliability estimation methods and demonstration tests are developed. It is recognized that reliability cannot easily be demonstrated by an acceptance test. Instead of a reliability figure (λ or MTBF $=1/\lambda$), contractual requirements are for a reliability assurance program. Maintainability, availability, and logistic support become important
1970–1980	Due to the increasing complexity and cost for maintenance of equipment and systems, the aspects of man-machine interface and life-cycle cost become important. Customers require demonstration of reliability and maintainability during the warranty period. Quality and reliability assurance activities are made project specific and carried out in close cooperation with all engineers involved in a project. Concepts like product assurance, cost effectiveness and systems engineering are introduced. Human reliability and product liability become important
1980–1990	Testability is required. Test and screening strategies are developed to reduce testing cost and warranty services. Because of the rapid progress in microelectronics, greater possibilities are available for redundant and fault tolerant structures. Software quality becomes important
After 1990	The necessity to further shorten the development time leads to the concept of concurrent engineering. Total quality management (TQM) appears as a refinement to quality assurance as used at the end of the seventies. RAMS is used for reliability, availability, maintainability and safety, reliability engineering for RAMS engineering

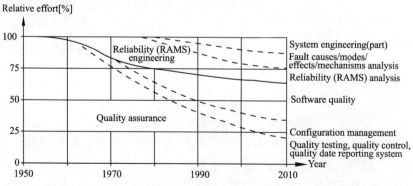

Figure 6.2 Approximate distribution of the effort between quality assurance and reliability (RAMS) engineering for complex equipment and systems with high quality and reliability (RAMS) requirements

Words and Expressions

mission [ˈmɪʃən] a task that is very difficult to complete	n. 任务
nonrepairable [ˌnɒnrɪˈpeərəbl] that cannot be repaired	adj. 不可修复的
entity [ˈentəti] that which is perceived or known or inferred to have its own distinct existence (living or nonliving)	n. 实体
time dependent [taɪm dɪˈpendənt]	时变；与时间有关的
transient [ˈtrænziənt] of a mental act; causing effects outside the mind	adj. 瞬时的
symptom [ˈsɪmptəm] (medicine) any sensation or change in bodily function that is experienced by a patient and is associated with a particular disease	n. 症状
creep [kriːp] someone unpleasantly strange or eccentric	n. 蠕变
intrinsic [ɪnˈtrɪnsɪk] belonging to a thing by its very nature	adj. 本质的；固有的
extrinsic [eksˈtrɪnsɪk] not forming an essential part of a thing or arising or originating from the outside	adj. 外在的；非固有的
deterministic [dɪˌtɜːmɪˈnɪstɪk] an inevitable consequence of antecedent sufficient causes	adj. 确定性的
bathtub curve [ˈbɑːtʌb kɜːv]	浴盆曲线
redundant [rɪˈdʌndənt] more than is needed, desired, or required	adj. 冗余的

Notes

[1] 本句可译为：现在的期望是，复杂设备和系统不仅在 $t=0$ 时（投入运行时）不存在缺陷和系统故障，而且在规定的时间间隔内能无故障地实现预期功能，并且在发生严重的甚至灾难性的失效时具有故障保护功能。

[2] 本句可译为：因此，维修分为预防性（或常规）维修，即以预定的间隔进行维修以

减少磨损故障，以及更正性（或非常规）维修，即在故障检测后进行维修，以使项目进入可再次执行所需功能的状态。

Questions for Discussion

1. What is the difference between the quality targets before and after the 1860s?
2. How to classify failures?
3. Give an example of one product which obeyed bathtub curve.
4. What is the difference between MTTF and MTBF?

6.2 Vehicle System Dynamics and Control

The railway train running along a track is one of the most complicated dynamical systems in engineering. Many bodies comprise the system and so it has many degrees of freedom. The bodies that make up the vehicle can be connected in various ways and a moving interface connects the vehicle with the track. This interface involves the complex geometry of the wheel tread and the rail head and nonconservative frictional forces generated by relative motion in the contact area.

The technology of this complex system rests on a long history. In the late 18th and early 19th century, development concentrated on the prime mover and the possibility of traction using adhesion. Strength of materials presented a major problem. Even though speeds were low, dynamic loads applied to the track were of concern and so the earliest vehicles used elements of suspension adopted from horse carriage practice. Above all, the problem of guidance was resolved by the almost universal adoption of the flanged wheel in the early 19th century, the result of empirical development, and dependent on engineering intuition.

Operation of the early vehicles led to verbal descriptions of their dynamic behavior, such as Stephenson's description of the kinematic oscillation. Later in the 19th century the first simple mathematical models of the action of the coned wheelset were introduced by Redtenbacher and Klingel, but they had virtually no impact on engineering practice. Actually, the balancing of the reciprocating masses of the steam locomotive assumed much greater importance.

A catastrophic bridge failure led to the first analytical model in 1849 of the interaction between vehicle and flexible track. The growing size of the steam locomotive increased the problem of the forces generated in negotiating curves, and in 1883 Mackenzie gave the first essentially correct description of curving. This became the basis of a standard calculation carried out in design offices throughout the era of the steam locomotive.

As train speeds increased, problems of ride quality, particularly in the lateral direction, became more important. The introduction of the electric locomotive at the end of the 19th century involved Carter, a mathematical electrical engineer, in the problem, with the result that a realistic model of the forces acting between wheel and rail was proposed and the first calculations of lateral stability carried out.

Generally, empirical engineering development was able to keep abreast of the requirements of ride quality and safety until the middle of the 20th century. Then, increasing speeds of trains and the greater potential risks arising from instability stimulated a more scientific approach to vehicle dynamics. Realistic calculations, supported by experiment, on which design decisions were based were achieved in the 1960s and as the power of the digital computer increased so did the scope of engineering calculations, leading to today's powerful modelling tools[1].

6.2.1 Coned Wheels

The conventional railway wheelset, which consists of two wheels mounted on a common axle, has a long history and evolved empirically. In the early days of the railways, speeds were low, and the objectives were the reduction of rolling resistance (so that the useful load that could be hauled by horses could be multiplied) and solving problems of strength and wear.

The flanged wheel running on a rail existed as early as the 17th century. The position of the flanges was on the inside, outside, or even on both sides of the wheels, and was still being debated in the 1820s. Wheels were normally fixed to the axle, although freely rotating wheels were sometimes used in order to reduce friction in curves. To start with, the play allowed between wheel flange and rail was minimal.

Coning was introduced partly to reduce the rubbing of the flange on the rail, and partly to ease the motion of the vehicle around curves. It is not known when coning of the wheel tread was first introduced. It would be natural to provide a smooth curve uniting the flange with the wheel tread, and wear of the tread would contribute to this. Moreover, once wheels were made of cast iron, taper was normal foundry practice. In the early 1830s the flangeway clearance was opened up to reduce the lateral forces between wheel and rail so that, typically, in current practice about 7 to 10 mm of lateral displacement is allowed before flange contact.

Coning of the wheel tread was well-established by 1821. George Stephenson in his "Observations on Edge and Tram Railways" stated that:

It must be understood the form of edge railway wheels are conical that is the outer is rather less than the inner diameter about 3/16 of an inch. Then from a small irregularity of the railway the wheels may be thrown a little to the right or a little to the left, when the former happens the right wheel will expose a larger and the left one a smaller diameter to the bearing surface of the rail which will cause the latter to lose ground of the former but at the same time in moving forward it gradually exposes a greater diameter to the rail while the right one on the contrary is gradually exposing a lesser which will cause it to lose ground of the left one but will regain it on its progress as has been described alternately gaining and losing ground of each other which will cause the wheels to proceed in an oscillatory but easy motion on the rails.

This is a very clear description of what is now called the kinematic oscillation (shown in Figure 6.3).

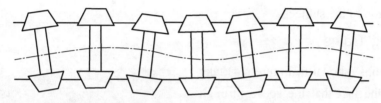

Figure 6.3   The kinematic oscillation of a wheelset

The rolling behavior of the wheelset suggests why it adopted its present form. If the flange is on the inside the conicity is positive and as the flange approaches the rail there will be a strong steering action tending to return the wheelset to the center of the track. If the flange is on the outside, the conicity is negative and the wheelset will simply run into the flange and remain in contact as the wheelset moves along the track. Moreover, consider motion in a sharp curve in which the wheelset is in flange contact. If the flange is on the inside, the lateral force applied by the rail to the leading wheelset is applied to the outer wheel and will be combined with an enhanced vertical load thus diminishing the risk of derailment. If the flange is on the outside, the lateral force applied by the rail is applied to the inner wheel, which has a reduced vertical load, and thus the risk of derailment is increased.

Guidance by railway tracks

As was explicitly stated by Brunel in 1838 it can be seen that for small displacements from the center of straight or slightly curved track the primary mode of guidance is conicity and it is on sharper curves, switches, and crossings that the flanges become the essential mode of guidance[2].

Lateral oscillations caused by coning were experienced from the early days of the railways. One solution to the oscillation problem that has been proposed from time to time, even down to modern times, was to fit wheels with cylindrical treads. However, in this case, if the wheels are rigidly mounted on the axle, very slight errors in parallelism would induce large lateral displacements that would be limited by flange contact. Thus, a wheelset with cylindrical treads tends to run in continuous flange contact.

In 1883 Klingel gave the first mathematical analysis of the kinematic oscillation and derived the relationship between the wavelength L and the wheelset conicity l, wheel radius r_0, and the lateral distance between contact points $2l$ as

$$A = 2\pi \left(\frac{r_0 l}{\lambda}\right)^{\frac{1}{2}} \tag{6.1}$$

Klingel's formula shows that as the speed is increased, so will the frequency of the kinematic oscillation. Any further aspects of the dynamical behavior of railway vehicles must be deduced from a consideration of the forces acting, and this had to wait for Carter's much later contribution to the subject.

6.2.2 Tilting Trains

Tilting train

Tilting trains (shown in Figure 6.4) take advantage of the fact that the speed through curves is principally limited by passenger comfort and not by either the lateral forces on the track or the risk of overturning, although these are constraints that cannot be ignored. Tilting the vehicle bodies inwards on curves reduces the acceleration experienced by the passenger, which permits higher speeds and provides a variety of operational benefits.

Figure 6.4 Tilting train (Series 283, JR[3] Hokkaido[4])

(1) Configurations

There are three basic mechanical configurations in tilting control, which are: tilt through or across the secondary suspension; tilt above the secondary suspension; and tilt below the secondary suspension. In the first two configurations, it is necessary to accommodate the increased deflections of the lateral secondary suspensions caused by the higher curving speeds, something that is not needed when tilt action is provided below the secondary suspensions.

Most tilting systems involve the use of an additional mechanism to achieve larger angles than are possible by controlling the roll suspension alone, although there have been some studies of limited angle tilting without using an additional mechanism, most recently in Germany and Japan.

(2) Control

The control objective is clearly to improve the vehicle response to the intended or deterministic track inputs (i.e. the curves), and a strong focus of control development has been to achieve this, without degrading the straight-track ride quality, i.e., the response to the unintended random track irregularity inputs. The maximum tilt angle is constrained by the mechanical design of the vehicle taking gauging issues into account, but an appropriate compensation factor must be selected to determine what cant deficiency the passengers should experience on a steady curve and what roll rate is acceptable on transitions. Those are clearly of primary importance to comfort on the steady curves and on curve transitions, and there are standard methods developed to predict the comfort levels.

Tilting is a low bandwidth control and, hence, presents a low demand in the response speed of the controller and the actuator. However, an accurate and timely measurement of the cant deficiency (or curvature and cant) is required to derive an appropriate command signal for the tilted angle as vehicles negotiate curves. It is essential to avoid both delays in tilting action on curve transitions and to minimize dynamic interactions with the lateral suspensions, which would lead to a reduced ride quality on straight track. A commonly used solution is a so called command-driven strategy, where the command signal is derived from a lateral accelerometer mounted on a non-tilting part of the bogie in the preceding vehicle with the filter designed such that its delay is compensated for by the preview or precedence effect.

Alternatively, the tilting action may be initiated by feeding the vehicle controllers with train positions just before its entering the curve with the help of a database, which defines the track, including the curve data. Accurate train positioning may be obtained with the use of track side beacons or more recently global positioning system (GPS) receivers.

(3) Implementation status and trends

Tilting is now an established and mature railway technology, with many successful, operational examples of tilting trains worldwide. The major changes have related to a significant progression towards using efficient and reliable electro-mechanical actuators instead of the hydraulic or pneumatic technology that dominated the early systems. For instance, a development by Sumitomo in Japan is a limited angle-tilt system, using the air springs alone—this is being used to give 2° of tilt for a new airport express in Nagoya as well as to provide a similar tilt angle as part of the FastTech360 development for the Shinkansen trains. Another concept that does not require a tilting bolster but which provides a somewhat larger tilt angle of around 4° was produced by Bombardier, during the 1990s, principally for regional rather than high speed trains—it employs an active anti-roll torsion bar and is now offered as an option for their FlexCompact bogie range.

Although there have been recent theoretical studies aimed at improving the response using local measurements on each vehicle, practical development of control laws has focused upon providing increasingly high quality information about the intended track alignment. The ultimate would be an entirely feed-forward system deriving information from a track database, but the accuracy and integrity requirements of such data are difficult, and so rather than developing this directly the industry approach is an evolution from the command-driven with precedence schemes, mentioned in the previous sub-section. The Alstom Tiltronix system still relies primarily upon its train sensors, but where there is good correlation of these signals to stored track data, it is able to provide enhanced performance.

A unsolved problem is the increased incidence of motion sickness that occurs in tilting trains—while there are a number of indications of the causes that have provided some help towards reducing the problem, particularly related to the passengers' sensitivity to roll velocity, an effective analytical understanding of the complex human interactions that might offer an optimized solution remains to be developed[5]. Nevertheless, it is clear that tilting has become a standard technology for many high speed trains.

6.2.3 Active Secondary Suspensions

In contrast to tilting, which improves the ride quality from the viewpoint of the deterministic curving or gradient inputs, active secondary suspensions are concerned with the stochastic track irregularity inputs.

(1) Configurations

Active control can be applied to any or all of the suspension degrees-of-freedom, but when applied in the lateral direction, it will implicitly include the yaw mode, and in the vertical direction will include the pitching mode.

A number of configurations for full active control are possible. Actuators can be used to replace the passive suspensions, and the suspension behavior will be completely controlled via active means. In practice, however, it is more beneficial that actuators are used in conjunction with passive components. When connected in parallel, the size of an actuator can be significantly reduced as the passive component will be largely responsible for providing a constant force to support the body mass of a vehicle in the vertical direction or quasi-static curving forces in the lateral direction. On the other hand, fitting a spring in series with the actuator helps with the high frequency problem caused by the lack of response in the actuator movement and control output at high-frequencies, and in practice, a combination of a parallel spring for load-carrying and a series spring to help with the high-frequency response is the most appropriate arrangement. The stiffness of the series spring depends upon the actuator technology—a relatively high value can be used for technologies such as hydraulics that have good high-frequency performance, and a softer value for other technologies for which achieving a high bandwidth is more problematic. Actuators can also be mounted between adjacent vehicles, although the improvement of ride quality is less significant.

(2) Control

The active control strategies have been extensively studied and many of the proposed schemes are based on the principle of "sky-hook" damping. Sky-hook damping is a high-bandwidth system, which can be used to give other improvements in suspension performance, largely through the provision of damping to an absolute datum. High levels of modal damping can be achieved without increasing the suspension's transmissibility at high frequencies. However, the control approach also creates large deflections at deterministic features such as curves and gradients. Although this can be accommodated in the control design, e.g., by filtering out the low frequency components from the measurements that is largely caused by track deterministic features, it should be recognized that reducing the deterministic deflections to an acceptable level will compromise the performance achievable with "pure" sky-hook damping.

Active secondary suspensions can also be used to provide a low bandwidth control, which is similar to tilting controls, in that the action is intended to respond principally to the low frequency deterministic track inputs. In low bandwidth systems, there will be passive elements which dictate the fundamental dynamic response, and active elements whose function is associated with some low frequency activity. A particular use of the concept is for maintaining the average position of the suspension in the center of its working space, thereby minimizing contact with the mechanical limits of travel, and enabling the possibility of a softer spring to be used. This is a powerful technique for the lateral suspensions because curving forces are large, and without centering action, there may sometimes be significant reductions in ride quality while curving.

(3) Implementation status trends

A great variety of options were studied during the 1990s, and following this work the world's first operational full-active lateral suspension has been in service since 2001 for JR East series E2 and E3 Shinkansen trains in Japan. The system, developed by Sumitomo, currently uses pneumatic actuators (shown in Figure 6.5) and has a sophisticated "H-infinity" modal controller (shown in Figure 6.6) that provides improved performance of the yaw and roll modes of the vehicle body. The system is being extended by using higher bandwidth electro-magnetic actuators for the Shinkansen Fastech 360-S program.

Active secondary suspension concepts have been well developed and the technology proven to a level where widespread introduction would be straightforward, but there is very limited use in service.

More advanced concepts involving control of multiple vehicle dynamic modes are possible, and at least one has been demonstrated experimentally for an innovative RailCab system being developed in Germany. This involves a technically complex, full-active suspension/tilt module that yields a comfortable suspension and a reduced sensitivity to track irregularities. It is, however, interesting to note that even simple control integration possibilities have not been seriously considered for conventional railway vehicles. For example, although tilting has been combined with a lateral centering device, full integration of tilt and active lateral suspension has not been studied in any detail, even though this might help to mitigate the problems of motion sickness. Another concept that has received very limited attention is that of an active suspension utilizing inter-vehicle actuators, an approach that potentially has a number of benefits—fewer actuators are needed, lower bandwidth actuator technology can be utilized and because they are not bogie-mounted they are in a more favorable vibration environment.

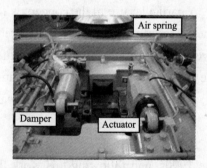

Figure 6.5　Pneumatic actuators for Sumitomo active lateral suspension

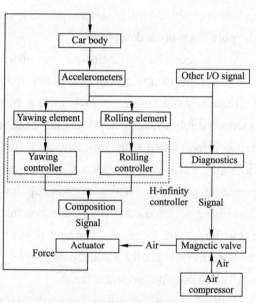

Figure 6.6　Control scheme for Sumitomo active lateral suspensions

Words and Expressions

nonconservative [ˌnɔnkən'sɜːvətɪv] not opposed to great or sudden social change	*adj.* 非保守的
reciprocating [rɪ'sɪprəkeɪtɪŋ] moving alternately backward and forward	*adj.* 往复的
empirical [ɪm'pɪrɪkl] derived from experiment and observation rather than theory	*adj.* 经验主义的

cone [kəʊn] to form cone-shaped artifact	*v.* 使成锥形
ride quality [raɪd ˈkwɒlɪti]	行驶质量
lateral [ˈlætərəl] situated at or extending to the side	*adj.* 侧面的；横向的
flangeway [flændʒweɪ] flange groove	*n.* 轮缘槽
clearance [ˈklɪərəns] the distance by which one thing clears another; the space between them	*n.* 空隙
kinematic oscillation [ˌkɪnɪˈmætɪk ˈɒsəˈleɪʃən]	[铁路] 蛇行运动
conical [ˈkɒnɪkl] relating to or resembling a cone	*adj.* 圆锥形的
conventional [kənˈvenʃənl] following accepted customs and proprieties	*adj.* 传统的；常见的；惯例的
mount on [maʊnt ˈɒn]	安装在……上面
flange [flændʒ] a projection used for strength or for attaching to another object	*n.* 轮缘
conicity [kɒˈnɪsɪti]	*n.* 锥度
tilting train [tɪltɪŋ ˈtreɪn]	摆式列车
secondary suspension [ˈsekəndəri səˈspenʃən]	二级悬挂
deflection [dɪˈflekʃən] a twist or aberration; especially a perverse or abnormal way of judging or acting	*n.* 变形量；歪斜
mature [məˈtʃʊr] develop and reach maturity; undergo maturation	*adj.* 成熟的
hydraulic [haɪˈdrɔːlɪk] moved or operated or effected by liquid (water or oil)	*adj.* 液压的
pneumatic [njuːˈmætɪk] of or relating to or using air (or a similar gas)	*adj.* 气动的
air spring [eə sprɪŋ]	空气弹簧
bogie [ˈbəʊgi] an evil spirit	*n.* 转向架
anti-roll torsion bar [ˈænti rəʊl ˈtɔːʃən ˈbɑː(r)]	抗侧滚扭杆

actuator [ˈæktʃueɪtə] a mechanism that puts something into automatic action	n. 执行机构

Notes

[1] 本句可译为：20世纪60年代实现了支持实验的现实计算，随着数字计算机功能的增加，工程计算的范围也随之增加，从而形成了当今强大的建模工具。

[2] 本句可译为：正如1838年Brunel明确指出的那样，可以看出，在偏离直线中心或稍微弯曲轨道中心的小位移上，主要的导向模式是（车轮）的锥度，而在更弯曲的曲线、道岔和交叉口上，则主要靠轮缘进行基本的制导。

[3] JR：日本铁路公司（Japan Railways的缩写）。

[4] JR Hokkaido：日本北海道铁路。

[5] 本句可译为：一个尚未解决的问题是，在倾斜列车中乘客晕车的概率会增加，尽管有许多成因解释可以减少该问题的发生，特别是与乘客对滚动速度的敏感性有关，但是对复杂的人类相互作用的有效分析的理解可能会提供一个优化的解决方案，然而，这仍然有待开发。

Questions for Discussion

1. Why is it important to control the vehicle dynamics?
2. Why do the wheels have to be coned?
3. Use your own words to explain the reason of kinematic oscillation.
4. What are the three basic mechanical configurations in tilting control?

6.3 Vehicle Vibration and Noise Control

Why do trains make so much noise

To some of us the sound of a passing train is music to the ears. Audio recordings of trains are sold; the sound of a steam engine laboring up a gradient or passing at speed may evoke a strong impression of its power or the nostalgia of a lost age. But to many people the noise from passing trains is unwanted and can be considered a disturbance.

It has always been so. The early railways were often subject to considerable opposition. The following was written in 1825, in a letter to the Leeds Intelligencer:

"Now judge, my friend, of my mortification, whilst I am sitting comfortably at breakfast with my family, enjoying the purity of the summer air, in a moment my dwelling ... is filled with dense smoke, ... Nothing is heard but the clanking iron, the blasphemous song, or the appalling curses of the directors of these infernal machines."

Nevertheless, although some objections such as this were attributed to environmental reasons including noise, most were based on economic or aesthetic arguments. An interesting example occurred as early as 1863, when the Manchester, Buxton, Matlock and Midlands Junction Railway (later to become part of the Midland Railway) in England was forced to build its line in a shallow cut-and-cover tunnel almost 1 km long so that it should not be visible from the Duke of Rutland's home at Haddon Hall[1]. Today such schemes and changes in alignment are not uncommon to mitigate noise, but the idea is clearly not new.

Particularly since the 1960s, environmental noise has become an increasingly important issue. Noise is often identified as a source of dissatisfaction with the living environment by residents. As environmental noise levels have increased, the population has become increasingly aware of noise as a potential issue. This applies to railway noise in common with many other forms of environmental noise. Opposition to new railway lines is now often focused on their potential noise impact. This may be because noise is quantifiable in a way that aesthetics are not, so that complaints about the railway as such become focused on the issue of noise. However, as noted by the Wilson Report of 1963, "There is a considerable amount of evidence that, as living standards rise, people are less likely to tolerate noise."

It was estimated in 1996 that 20% of the population of Western Europe lived in areas where the ambient noise level1 was over 65 dB and as many as 60% in areas where the noise level was over 55 dB. The major source of this noise is road traffic, which accounts for around 90% of the population exposed to levels of noise over 65 dB (i.e. 18 of the 20%). However, railways and aircraft are also important sources of noise in the community. Rail traffic accounts for noise levels over 65 dB for 1.7% of the population.

While not everyone reacts to noise in the same way, it is no surprise that in terms of annoyance, between 20 and 25% of the population are annoyed by road traffic noise and between 2 and 4% by railway noise. Nevertheless it has been found that for the same level of noise, railways are less annoying than road traffic, leading to a "railway bonus" in some national standards, recommendations and guidelines, notably a "bonus" of 5 dB in Germany.

The prevalence of high-noise levels and increased public awareness of noise has led to the introduction of legislation to limit sound levels, both at receiver locations and at the source. European legislation has existed since the 1970s to control the sound emitted by road vehicles and aircraft. For road vehicles, reductions in the levels obtained during the drive-by test of 8~11 dB have been achieved between 1973 and 1996. However, it is widely recognized that this does not translate into equivalent reductions of noise in traffic, due to a mismatch between the test conditions and typical traffic conditions. In the former (low speed acceleration under high-engine speed) engine noise dominates whereas in traffic usually tyre noise dominates for speeds of 50 km/h and above. Changes in the test procedure are proposed to overcome this. Rubber tyres are clearly not quiet, being responsible for much of the noise exposure due to transport.

By contrast aircraft noise has been reduced by stricter noise certification and by night-time flying bans and other operational measures. The introduction of high bypass ratio turbo-fan engines has reduced noise by 20~30 dB since the early 1970s, although the rapid increase in the number of flights means that the noise exposure has in many situations close to airports continued to rise or at least remained steady.

For railway noise, the difficulty of separating the influence of track and vehicle and the consequent difficulty of defining a unique source value for a particular vehicle have contributed to the long delay in the introduction of source limits. Legal limits on the noise emitted by individual rail vehicles have only been introduced in Europe since 2002. These have been achieved through the means of "Technical Specifications for Interoperability" (TSI), which are intended primarily to allow interoperability of vehicles between different countries in Europe. Such limits have the potential to reduce railway noise in the long term.

Noise limits at receiver locations apply in many countries. These were mostly introduced initially to apply to new lines or altered situations, providing for mitigation such as noise barriers or secondary glazing where limits were exceeded. However, the introduction of the Environmental Noise Directive (END) has led to the requirement to produce noise maps of existing sources and to develop Action Plans to reduce noise in identified "hot spots". These, too, will mean that railway operators and infrastructure companies will have to consider how to minimize noise.

As well as noise, vibration from railways can cause annoyance. This may be due to feelable vibration, usually in the range 2 to 80 Hz, or due to the radiation of low frequency sound transmitted through the ground, usually in the range 30 to 250 Hz. Vibration may also cause objects to rattle, adding to the sensation.

Noise at a particular receiver location can be reduced by secondary measures, either in the transmission path such as noise barriers or at the receiver such as by installing windows with improved acoustic insulation. To a lesser extent vibration can also be dealt with by secondary measures, such as mounting sensitive buildings on isolation springs. Nevertheless, reduction of noise and vibration at the source is generally more cost effective. On the other hand, it is also generally true that noise control at source is more complex, as it requires a good knowledge of the mechanisms operating within the source. It is important that safe and economic operation of the equipment, in this case the railway, is not impeded by changes aimed at reducing noise. The railway is often seen as a conservative industry where there is reluctance to change the way things are done, particularly because of potential implications for safety or operational efficiency.

Railways are generally acknowledged to be an environmentally-friendly means of transport with the potential to operate with considerably less pollution, energy use and CO_2 emissions per passenger-km than road or air. High speed trains have been found to compete effectively with air transport on routes up to 3 hours or more, achieving large market shares on routes. Mass transit systems hold the key to urban mobility. In order to improve the market share of rail transport, and thereby improve sustainability, it is imperative that noise is reduced.

6.3.1 Needs for a Systematic Approach to Noise Control

The first step in noise control is to identify the dominant source. There are many different sources of noise from a railway, and in different situations the dominant source may vary. Notably on North American freight railways, a major issue for environmental noise is related to locomotive warning signals. It is obligatory to sound the horn in an extended sequence on the approach to road crossings. There are many such crossings, especially in populated areas and so it is a major source of annoyance, particularly from operations at night. In other situations, such as stations in urban areas, the public address system may be the major source of noise in the immediate neighborhood. However, the most important source of noise from railway operations is usually rolling noise caused by the interaction of wheel and rail during running on straight track. Other sources include curve squeal, bridge noise, traction noise and aerodynamic noise. Noise inside the vehicle also includes all of these sources, as well as others such as air-conditioning fan noise.

Having identified the dominant source, the next step is to quantify the various paths or contributions. Focusing on exterior rolling noise, the vibration of the wheel and the rail can be identified as potential sources. Early attempts to understand the problem tended to be polarized into attributing the noise solely to one or the other. More recently, however, it has become widely recognized that both wheels and rails usually form important sources which make similar contributions to the overall sound level. Prediction models allow their relative contributions to be quantified (measurement methods can also be used). Clearly, effective noise control requires both sources to be tackled. For example, in a situation where wheel and rail contribute equally to the overall level, a reduction of 10 dB in one of them, while the other is unchanged, will produce a reduction of only 2.5 dB in the total.

The next step is to understand how each source can be influenced. Here, the theoretical models allow the sensitivity of the noise to various design parameters to be investigated (measurements alone do not). Noise control principles can be considered in terms of reduced excitation, increased damping, vibration isolation, acoustic shielding or absorption.

From these principles, actual designs can be developed and tested, first using the prediction model, then in laboratory tests and ultimately in practical tests on the operational railway. Tests should be carried out in a controlled situation; where possible not just the noise but intermediate parameters such as vibration should be measured.

6.3.2 Sources of Railway Noise and Vibration

There are many sources of noise and vibration in the railway system. The main ones are summarized below. The dominant source at most speeds is rolling noise which increases with train speed V at a rate of about 30 log10 V, i.e. a 9 dB increase in sound level for a doubling of speed. Traction noise is much less dependent on train speed so that it is often dominant only at low speeds. Conversely, aerodynamic noise has a much greater speed dependence than other sources and so becomes dominant at high speeds.

The audible frequency range extends from 20 to 20,000 Hz and this broadly defines the range of interest for acoustic analysis. Within this range some sources, such as bridge noise or ground-borne noise are concentrated at low frequencies, while squeal nose can occur at very high frequencies.

(1) Rolling noise

The most important source of noise from railways is rolling noise caused by wheel and rail vibrations induced at the wheel/rail contact. Roughness on the wheel and rail running surfaces induces vertical vibration of the wheel and rail systems according to their dynamic properties. Figure 6.7 indicates this in the form of a flow chart, while Figure 6.8 shows the mechanism visually. The main wavelengths of roughness that are relevant to rolling noise are between about 5 and 500 mm. This vibration is transmitted into the wheel and track structures leading to sound radiation. Often, both wheel and track vibration are important to the overall noise level and both must be taken into account. Rolling noise is fairly broad-band in nature, the relative importance of higher frequency components increasing as the train speed increases.

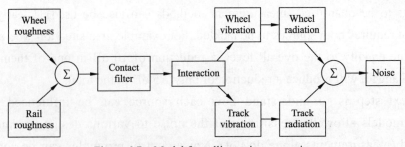

Figure 6.7 Model for rolling noise generation

Another complication is that the vibrations of both wheel and rail are induced by the combination of their roughnesses. A typical situation is that wheels fitted with cast-iron brake blocks have large roughness with wavelengths around 40~80 mm. For a train speed of 100 km/h this roughness excites frequencies where the track vibration radiates most sound. In such a situation, is the vehicle or the track responsible for the noise? Clearly, both are.

Compared with the other sources, rolling noise has been the subject of the greatest amount of research over the years, and will be treated in most detail in the first half of the book. Impact noise is excited by irregularities such as wheel flats, rail joints, dipped welds or switches and crossings. This has similarities to rolling noise as the excitation is primarily vertical, but it differs in that non-linearity in the contact zone is important and cannot be neglected.

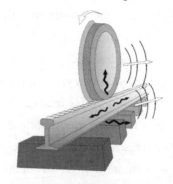

Figure 6.8 Illustration of the mechanism of generation of rolling noise

(2) Curve squeal

Curve squeal noise is also caused by the interaction between wheel and rail but has a quite different character. It is strongly tonal, being associated with the vibration of the wheel in one of its resonances. This is excited by unsteady transverse forces at the contact occurring during curving. It is also necessary to distinguish between squeal caused by lateral creepage, or "top-of-rail" squeal, and "flange squeal" or "flanging noise". The latter has a more intermittent nature and generally has a higher frequency content, consisting also of many more harmonics rather than a single dominant tone. A similar phenomenon is brake squeal in which tonal or multi-tonal noise is emitted during braking.

(3) Bridge noise

When a train runs over a bridge the noise emitted can increase considerably, depending on the type of bridge. The bridge is excited by dynamic forces acting on it from the track. Bridges vary greatly in construction. Steel bridges with direct fastenings are usually the noisiest and can be more than 10 dB noisier than plain ballasted track, in some cases up to 20 dB. Moreover, the increase in noise on a bridge is usually greatest at low frequency, so that the A-weighted sound level does not fully take account of this effect.

(4) Aerodynamic noise

In contrast to most other railway noise sources, which are caused by the radiation of sound by the vibration of solid structures, aerodynamic noise is caused by unsteady air flow over the train. Aerodynamic sources of noise generally increase much more rapidly with speed than mechanical sources. Both broad-band and tonal noise can be generated by air flow over various parts of the train, but much of the sound energy is concentrated in the lower part of the frequency region. Considerable understanding of the sources of aerodynamic noise has been obtained in recent years but modelling is much more involved than for vibroacoustic problems and the models are still at a relatively early stage.

(5) Ground vibration and noise

As noted already, vibration transmitted through the ground can be experienced in two ways. Low frequency vibration between about 2 and 80 Hz is perceived as feelable "whole body" vibration. This tends to be associated most with heavy freight trains at particular sites. Higher frequency ground-borne vibration from about 30 to 250 Hz causes the walls, floors and ceilings of rooms to vibrate and radiate low frequency noise. This is associated more with trains in tunnels in urban areas such as metro operations but can also be significant for surface railways, particularly where noise barriers block out the direct airborne sound.

Whereas air is essentially a uniform medium, the ground can have a layered structure and its properties vary considerably from one site to another and even within a single site. Thus prediction of absolute levels of vibration and ground-borne noise is extremely difficult, requiring complex models of the ground and in turn detailed information about ground properties, as well as models of the train and track and possibly also the buildings at the receiver location.

(6) Internal noise and vibration

Many of the sources that are relevant to environmental noise also lead to noise inside the railway vehicle. However, their frequency content is modified by the transmission paths into the vehicle, which may be both structure-borne and airborne. Consequently noise spectra inside vehicles tend to be dominated by low frequency sound, as is also the case for automobiles. In addition, low frequency vibration inside the vehicle affects ride comfort.

(7) Other sources of railway noise

Other sources of railway noise include:

- Traction noise from diesel engines, exhaust and intake, from traction motors and fans, from gearboxes, turbochargers, etc.;
- Warning signals from trains (horns, etc.) and fixed installations (i.e. level crossings);
- Track maintenance equipment;
- Shunting noise, in particular the noise from impacts between vehicles.

6.3.3 Rail Traffic Noise and Vibration Mitigation Measures in Urban Areas

Studies of noise and vibrations caused by traffic operations on tracks, as a part of analysis of railway transportation impact on the environment, are often considered as one and the same discipline because both phenomena have many common physical characteristics. They are both analyzed as a wave phenomenon: noise is defined as sound waves propagating through the air, while vibrations travel through the ground also in the form of waves. They are both result of vibrations of wheels and rails during vehicles rolling on track i.e. dynamic forces arising due to wheel-rail interface roughness. At high frequencies, this excitation energy expands through the air in the form of sound waves (noise), whereas lower frequency waves transmit from the rails to the lower parts of the track structure, through the ground and to the objects in the ground (shown in Figure 6.9). Roughly speaking, vibrations and structure-born noise occur in the frequency range 0~100 Hz and noise 30~2,000 Hz.

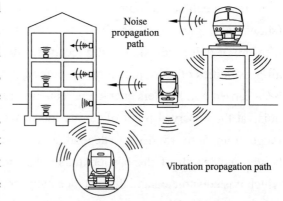

Figure 6.9 Rail traffic noise and vibratio ns

There are four main groups of rail traffic noise and vibration mitigation measures:
- Reduction at source;
- Reduction of propagation;
- Isolation of receiver;
- Economic measures and regulations.

The first group represents the so-called primary measures, while the other three groups are considered to be secondary measures of protection against rail noise and vibration. This chapter will consider only the first two measures that are primarily related to the track and rail vehicles.

(1) Reducing noise and vibrations at source

Reduction of noise and vibration at source can be achieved by:
- Increasing the elasticity of the track superstructure;
- Eliminating the running surface discontinuities;
- Regular maintenance of the rail running surface;
- Regular wheel re-profiling;
- Selecting the appropriate type of rail vehicle;
- Reducing the speed of rail vehicles.

(2) Noise and vibration propagation mitigation

Increasing the distance between the track and the receiver is one of the most effective measures of noise and vibration reduction, but only appropriate in cases where the cost of land purchase is less than the estimated cost of the implementation of other mitigation measures. At a distance of 500 m from the rail track people no longer perceive the rail traffic vibrations, while the air borne noise decreases by about 20 dB(A).

Implementation of noise protection barriers (shown in Figure 6.10) is the most effective measure of noise reduction at the receiver, after all possibilities of noise reduction at source have been carried out. Barriers enable the reduction of noise at the observed point of immission in the range from 5 to 15 dB(A), depending on the height and length of the barrier, material from which it is constructed and the distance between the source and receiver.

Figure 6.10 Noise protection barrier

Use of trenches to control ground-borne vibration is analogous to controlling airborne noise with sound barriers. A trench can be a practical method for controlling rail traffic vibration from at-grade track. Studies have shown that the reduction in vibration at frequencies of 30 Hz requires a minimum trench depth of 5 m.

By performing various methods of soil stiffening (lime modification, lime injection and jet grouting) it is possible to get soil with a larger vibration absorption capacity. Research has shown that reducing the coherence of the soil allows reducing vibration by 14 dB within the frequency of 4 to 32 Hz.

A convenient alternative solution to reduce the propagation of vibration in urban areas, where performance of trenches is impossible due to the existing development, is the performance of the underground barriers near the rail track[2]. Such barriers are performed in situ, by mixing existing soil with live lime or cement, in the form of pillars with diameter from 0.5 to 1.0 m. The depth of the barriers depends on the frequency of vibration that needs to be reduced. The lower the frequency of vibration, the greater is the needed depth of such barriers (depth of the barrier needed to reduce the vibration for about 25% ranges from 10 to 15 m).

By placing the rail track in tunnels, the greatest effect of noise and vibration reduction can be achieved. With additional application of specially designed tracks in tunnels, such as "floating slab tracks", it is possible to achieve vibration reduction for up to 40 dB. This measure is rarely used exclusively to reduce rail traffic noise and vibration, since the cost of tunnel construction, maintenance, lighting and ventilation is very high.

Therefore, the reasons for lowering the rail tracks in urban areas in the underground are, in the first place, operational in nature, which can be confirmed by the example of the town of Split. During 1970s, single track divided the town into two parts causing traffic jams on level crossings and creating obstacles for normal life of the inhabitants. Therefore, the plan was made for cutting and covering of railway and for substituting single track on ground with electrified double track in tunnel. By completion of the tunnel in 1979, both traffic congestion and pollution of the surrounding residential areas by rail traffic noise were completely eliminated.

Words and Expressions

labor up a gradient [ˈleɪbə ʌp ə ˈɡreɪdiənt]	爬坡
nostalgia [nɒˈstældʒə] longing for something past	n. 乡愁；怀旧之情
disturbance [dɪˈstɜːbəns] activity that is an intrusion or interruption	n. 干扰
tyre [ˈtaɪə(r)] hoop that covers a wheel	n.（橡胶）轮胎
hot spot [hɒt spɒt]	热点
feelable vibration [ˈfiːləbl vaɪˈbreɪʃən]	能感觉到的振动
flow chart [fləʊ tʃɑːt]	流程图
creepage [ˈkriːpɪdʒ]	n.（材料、金属等的）蠕变
cast-iron brake block [kɑːst aɪən breɪk ˈblɒk]	铸铁制动块
squeal [skwiːl] a high-pitched howl	n. 尖叫声
fastening [ˈfɑːsnɪŋ] restraint that attaches to something or holds something in place	n. 紧固零件
A-weighted sound level [ˈweɪtɪd saʊnd ˈlevl]	计权声级 A
vibro acoustic [ˈvaɪbrə əˈkuːstɪk]	声振
noise and vibration mitigation [nɔɪz ænd vaɪˈbreɪʃən ˌmɪtɪˈɡeɪʃn]	降噪减振
point of immission [pɔɪnt əv ɪˈmɪʃən]	输入点
soil stiffening [sɔɪl ˈstɪfnɪŋ]	土壤硬化
lime modification [laɪm ˌmɒdɪfɪˈkeɪʃn]	石灰改性
lime injection [laɪm ɪnˈdʒekʃən]	石灰粉喷射
jet grout [dʒet ɡraʊt]	喷射灌浆
implementation [ˌɪmplɪmenˈteɪʃən] the act of accomplishing some aim or executing some order	n. 安装启用
electrified double track [ɪˈlektrɪfaɪd ˈdʌbl træk]	电气化复线铁道

Notes

[1] 本句可译为：早在 1863 年就出现了一个有趣的例子，当时英国的曼彻斯特、巴克斯顿、马特洛克和米德兰枢纽铁路（后来成为米德兰铁路的一部分）被迫在一条长约 1 千米的浅埋暗挖隧道中修建线路，以便让 Rutland 公爵从在哈顿庄园的家里看不到铁路。

[2] 本句可译为：在城市地区减少振动传播的一个方便替代解决方案是在轨道附近设置地下屏障，这特别适合随着城市发展很难挖沟的地方。

Questions for Discussion

1. List the differences of noise source between road vehicle, aircraft and rail traffic.
2. Why is it imperative that noise must have to be reduced?
3. Briefly describe the steps of noise control.
4. Briefly describe the noise and vibration mitigation measures.
5. What factors affect the noise mitigation effect of the barrier?

6.4　Vehicle Digital Design

Computer aided design (CAD) technology has changed the design means and the design concepts in almost all fields, whose development and application level has become one of the important symbols to measure the modernization of national science and technology and the industry[1]. This technology, after decades of research and being applied widely, has tended to be mature. The development of CAD technology from the two-dimensional drawing and 3D line frame model to the free curve and surface generation algorithm and surface modeling theory in the 1970s, and then to solid modeling theory and geometric modeling method in the 1980s, but now the development of CAD technology has integrated modeling, analysis, simulation methods. For the future of CAD technology's development, one of the views was mentioned that it would move to the direction of integration, networking, parameterization, standardization, specialization, openness, virtualization and intellectualization. High speed train, integrated modern machinery, materials, computer, electrical and other disciplines of high technology, was the epitome of the modern industrial development level in today's city. By the end of 2017, China has had the largest high speed railway network, the most complex high speed railway running environment and the most massive high speed railway passengers all over the world. At the same time, EMU products constituted the product spectrum that covered different speed grade, different organization form, and different products which were suitable for different environment. It was undoubted that the railway vehicle industry put forward a very high request. It was difficult to design and manufacture the appropriate railway vehicle products just only to rely on the traditional design and manufacturing method. It also needs to rely on CAD technology.

6.4.1 Current Situation of CAD Technology Used for Rail Vehicles in China

Rail vehicle using CAD technology started relatively late, which began to be paid attention from the 1970s. As early as in the 1980s, China Railway Corporation, according to the relevant technical policies drawn up by the national and the Ministry of Railways (named at that time), listed the computer aided design as one of the three major systems of computer application in the industry. Designers repeatedly proposed and emphasized strengthening the combination of product designing, processing and manufacturing, and invested actively the application of aided design and manufacturing in different conference on Computer Applications. To effectively promote its development, in 1987, Chinese Railway Corporation set up a special group for the application of CAD achievement appraisal. In the early 1990s, the means of simulation analysis which most of the domestic vehicle factory and the institute held covered all aspects basically such as the structural strength of components, vehicle or component fatigue, the train dynamics, vibration and noise and so on. China National Railway Locomotive & Rolling Stock Industry Corporation (LORIC) introduced ANSYS analysis software on a large scale for the first time guided and trained professional talent in 1999. At that time, the main development was complete autonomy. Then, the development of CAD began to advance such as the two times of development of AutoCAD. In 2000, a system development based on CAD software platform was proposed, which was between independent development and two times of development, and it was nearer than the two times development from the core layer, which was characterized by short development cycle, system stability and powerful, etc. In 2006, China Northern Locomotive and Rolling Stock Industry Group Corporation (CNR Group) held a product R & D (Research and Development) information work and the virtual product development technology training courses in Shanghai, officially launched the project development platform construction vehicle virtual product, which prompted the CAD technology to a new level. Vehicle virtual product development platform for the implementation of construction projects was to fully exploit the investment opportunities of treasury bonds, to ensure the hardware and software investment of CAD/CAE/CAM/ CAPP/PDM, and to expand vehicle virtual prototype development and application of systems engineering achievements. With the UGS, UFC in depth communication, the "China Railway Locomotive & Rolling Stock Industry 3D Acceleration Plan" was proposed.

Light rail transit track design and analysis—CAD

The development of the country in terms of rail vehicles, although it started late, developed very rapidly. The development of rail vehicles and CAD technology were closely connected with each other. Especially from the introduction, absorption, improvement and innovation of high speed train technology, to the "Harmony" EMU which were independent research and development, China high speed train technology had developed rapidly. In this period, with the rapid development of CAD technology, more vehicle technical personnel transferred to the field of CAD. State Key Laboratory of scientific research institutions and universities, workers were making full use of computer aided design technology, to shorten the gap with foreign countries by their efforts. While learning from foreign advanced EMU technology, they always took a new road of independent research and development. Changchun Railway Vehicles Co., Ltd. in 2007 combined the classical vibration theory, rigid body motion analysis theory and multi-body dynamics software SIMPACK for system simulation, and obtained consistent data to research the characteristic of the suspension system.

6.4.2 Finite Element Method (FEM)

The finite element method (FEM) is a numerical technique that is used to reduce a modelling problem down into small elements. These small elements are modelled as simple modules with external inputs from adjacent elements. FEM can be used to solve complex systems with good accuracy. FEM models consist of finite elements connected to other finite elements at points called nodes. Types of FEM that can be used in railway applications include modelling the rail vehicle structure, rail, air brake system, heat transfer due to braking and movement of fluids in tank wagons.

Modelling of the rail vehicle structure can be handled in two ways. For typical dynamic modelling, the vehicle structure can be considered either as a rigid body or as a single stiffness. This allows simplified longitudinal train and vehicle dynamics modelling and determination of the forces on the rail vehicle. The external forces can then be applied to a detailed FEM model of the rail vehicle body/frame to determine the expected stresses and fatigue lives. The FEM mesh of the structure is sized according to the accuracy desired, the loading type and the computational speed required. Figure 6.11 shows an FEM model of part of a rail vehicle frame modelling an anti-ride-up device. There are numerous structural FEM software packages on the market that can be used in the analysis, and the choice is based largely on ease of use, support available, accuracy and the pricing structure. FEM packages may also provide a way to determine the simplified vehicle stiffnesses that can be used in the dynamic analysis by either a dynamic train model or a detailed rail vehicle dynamic model.

The second method of modelling the rail vehicle structure is to use software that incorporates both an FEM of the vehicle as well as the dynamic model. This approach is very similar to the first approach, but it is simplified as the inputs and outputs of the FEM and dynamic models are handled by the software. This type of modelling is particularly suited to crash simulations where the deflection of the rail vehicle body and frame affect the vehicle dynamics. To reduce the computational effort required, the FEM models integrated with dynamic modelling software are generally simplified. As a result, detailed FEM rail vehicle body analysis typically uses standalone packages to increase the accuracy of the results. Models integrating vehicle dynamics and FEM can still be essential to get more accurate estimated forces where forces are modified by flexure or deformation in solid body components.

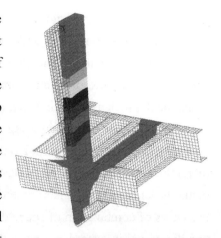

Figure 6.11   FEM of rail vehicle frame and anti-ride-up device

6.4.3   Key Technologies of Digital Design Direction of Rail Vehicles

(1) Integrated

The integrated design of products reflect three main points. First, collaborative design is distributed. This is to say, different designers will design a physical map in different places by the same software. Second, the complementary advantages of all kinds of design software are to realize the integration of data seamless connection. Third is about the integration of CAD/CAM. At present, the foreign mainly focus their attention and efforts on the aspects of CAD/CAE/CAM integrated design.

Taking axle box design of rail vehicles as an example, designers can rely on the operation function of design software to complete the drawing box in different places. The current software are different and there is no uniform standard, so each software is very mature just in some aspect, which causes the data transmission problems between different software. For example, the crankcase is imported, which is drawn out by the SolidWorks software into ANSYS, which need transform data format. So it is necessary to realize the seamless connection between software.

(2) Modular

There are many types of vehicles, so do its affiliated parts. A common bus can use several kinds of bogies, so as the same wheels share different bogie wheels, and the same chassis in the same series can be used for a variety of body bearing structure. The thought of modular design and manufacture is just put forward to solve these kinds of problems. The main idea of modular design is to divide the vehicle into several independent modules, and subdivided into sub-modules under the module. These modules can be composed of different kinds of vehicles with organically combining. This modeling method has stranger organized, and solves complex problems more easily. Under the sub-module level are some basic components, such as bus windows, doors, wheels, box axle and a series of spring. Therefore, the vehicles can be divided into segmentation modules, sub-modules, components, spare parts and so on. The task is to find out the relationship between the upper and lower levels, hierarchical relationships, relevance, the effects of breadth and depth and manufacturing methods, and then to create a three-dimensional CAD environment as the core database, with the views of combination of space design, and integrate the entire design and manufacture[2]. For the bogie, in accordance with the modular approach, can be mainly divided into many sub-modules, such as axle, wheel, axle box, inner and outer ring spring, bolster and side frame. The axle box can also be divided into the box, a series of springs and other key components of the module. These modules with certain characteristic parameters form different types of bogies by using the background rich databases resources.

(3) Parametric

The products designed, parts of their structure and mathematical models are also relatively fixed, and differences generally exist only in the structure size. In R & D, design process, in order to find the optimal results, the product needs to be revised many times. For such products, basic parameters and known conditions can be replaced by the corresponding variables. By calculation or query, automatic generation of graphic design it needs, the realization of this course is what we call the product parametric design. Parametric design is one of the most important research fields in CAD. Parametric technology allows designers to drive the geometric model of the products and parts through the design parameters. Compared with the traditional modeling methods, parametric design frees the designer from the trivial to piece together geometric elements of operation, which greatly simplifies the generation and the operation of modification of part model, improving the design efficiency. But how to get the appropriate parameters is a complex process. Rail vehicle itself is a complex system with complex structure, huge system, and high integrated. And in such a country as China, it has a vast territory, and there is diversity in national and geographical conditions and climatic environment, so there are different requirements of technical indexes for the different vehicle. At the same time, design parameters and technical indicators of the vehicle generally have polyphyly, hierarchical, transitivity, coupling characteristics. So to realize the design parameters of the vehicle, we need to find out the correlation between the parameters and the mapping relationship between parameters and technical indicators. Based on the construction of vehicles parameters design system, the forward vehicle design scheme can be realized.

(4) Database management and control

Vehicle design data are stored in the databases of various types of design information, as well as all the various types of locomotives parts, informative, varieties, complex relationships. To efficiently manage documents, we can reduce the waste of resources, shorten development cycles and share resources. So, we need to classify the data which are relevant, similarity, data versatility (replace) and other characteristics reasonably. Faced with such an era of big data, more than a dozen of academicians recommend to senior that China's national strategy should be developed for large data and development goals, making the development of top-level design principles, the key technology. In the high speed train, the vehicle top-level design plan is started. By means of networking, cloud computing, mobile Internet and other technologies, collaborative innovation, build parameters of the system platform.

6.4.4　Development Trend of CAD

For the design concept, it introduced new innovation theory for TRIZ (theory of inventive problem solving), universal design and reconfigurable design. Many scholars did a lot of researches about this theory in innovation design. For example, they put forward to realize product creative design based on the integrated of QFD (quality function deployment), CBR(case based reasoning) and TRIZ, and infected a mapping process of product design based on general design theory just was a knowledge-intensive program. Which indicated intelligent CAD with complex knowledge representation and reasoning would become an important direction of development of today's CAD. Fuzzy reconfigurable design principles and methods of mechanical products gave an answer that how to overcome the problems of resource reuse and achieve a rapid design of new products. The application of these design theories and achievements can be seen, to achieve the unique design of the product requires a corresponding theoretical support.

For the design method, it is going to change from reverse engineering to top-down design for rail vehicles designed. Because the old method that gave initial value before designing and repeated verification is consuming time and laborious. In contrast, the top-down design can avoid the repeated design, reduce the number of alternation and shorten the development cycle. In the process of realizing the top-down design, clearing up the parameters correlation of the vehicle and subsystem, subsystem and subsystem, and pointing out the mapping relationship between the parameters and technical requirements are the primary task.

For the design platform development, it would better realize the organic combination of modular and integrated. Modular and integrated both are necessary requirements for the development of digital design locomotive. Actually, they have been made a lot of achievements so far. However, the train belonging to complex electromechanical systems, involving multiple disciplines, is also divided into several departments even though in the design and manufacturing enterprises. Although existing subsystems have gotten a good application in department, information silos and data not unified were serious. Parts of integrated system have realized the unified management of data, but not realized professional design for subsystems. So design effects and design efficiency are not up to expectations.

For data management, it should accelerate basic technology research for the unstructured data processing technology, non-relational database management technology and visualization technology. If do that, it will be beneficial to integrate innovation resources and build the big data system platform with the help of cloud computing, Internet of things, and mobile Internet technology. Further, it also provides technical support system for intelligent design.

Words and Expressions

appraisal [əˈpreɪzl] the classification of someone or something with respect to its worth	*n.* 评估
domestic [dəˈmestɪk] of or inside a particular country	*adj.* 国内的
China National Railway Locomotive & Rolling Stock Industry Corporation (LORIC)	中国铁路机车车辆工业总公司，简称中车公司或 LORIC
China Northern Locomotive and Rolling Stock Industry Group Corporation (CNR Group)	中国北方机车车辆工业集团公司，简称 CNR Group
R&D (research and development) [rɪˈsɜːtʃ ænd dɪˈveləpmənt]	技术研发（在外资企业中特指研发部）
Changchun Railway Vehicles Co., Ltd.	中车长春轨道客车股份有限公司
integrate [ˈɪntɪɡreɪt] make into a whole or make part of a whole	*v.* 一体化
affiliated [əˈfɪlieɪtɪd] being joined in close association	*adj.* 附属的
chassis [ˈʃæsi] alternative names for the body of a human being	*n.* 底架
modular design [ˈmɒdjələ dɪˈzaɪn]	模块化设计
parametric [ˌpærəˈmetrɪk] of or relating to or in terms of a parameter	*adj.* 参数化的

polyphyly [ˈpɒlɪfɪli] a polyphyletic (Greek for "of many races") group is one characterized by one or more homoplasies: character states which have converged or reverted so as to appear to be the same but which have not been inherited from common ancestors	*n.* 多源性
hierarchical [ˌhaɪəˈrɑːrkɪkl] classified according to various criteria into successive levels or layers	*adj.* 分层的；等级体系的
TRIZ (theory of inventive problem solving) [ˈθɪəri əv ɪnˈventɪv ˈprɒbləm ˈsɒlvɪŋ]	发明问题解决理论
universal design [ˌjuːnɪˈvɜːsl ˈdɪˈzaɪn]	通用设计
reconfigurable [rɪˈkɒnfɪɡərəbl] can be make changes to the way	*adj.* 可重构的
fuzzy reconfigurable design principle [ˈfʌzi rɪˈkɒnfɪɡərəbl ˈdɪˈzaɪn prɪnsəpl]	模糊可重构设计原理

Notes

[1] 本句可译为：计算机辅助设计（CAD）技术改变了几乎所有领域的设计手段和设计理念，其发展和应用水平已成为衡量国家科技和工业现代化的重要标志之一。

[2] 本句可译为：其任务是找出上下层之间关系、分层关系、相关性、广度和深度的影响以及制造方法，然后以空间设计的组合视图，创建一个以三维 CAD 环境为核心的数据库，并把整个设计和制造结合起来。

Questions for Discussion

1. What are the characteristics of CAD technology development?
2. List an outline of CAD technology development for EMU in China.
3. What is the difference between the two ways of modelling of the rail vehicle structure?
4. What are the advantages of modular design?
5. Why are database management and control important in the process of vehicle digital design?

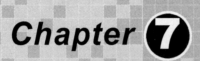

Chapter 7 Materials Science and Technology

7.1 Materials for Rail Traffic

7.1.1 Introduction

(1) Introduction of materials

Introduction of materials

Material refers to the matter by which human beings can process various kinds of products. It is the survival and development foundation of human beings. Since the 1970s, people regard the information, material and energy as the three pillars of contemporary civilization. In the 21st century, the new materials, information technology and biotechnology are also considered as the important symbols of new technological revolution. Accordingly, material plays an important role in the national economy, national defense and people's life and has made great contributions to the human civilization and the progress of the world.

The materials in nature are rich and colorful. Thus, their classification methods are varied. According to purpose, the materials can be divided into the structural materials, functional materials, etc. For the structural materials, the mechanical and processing properties are emphasized. For the functional materials, the physical and chemical properties are emphasized.

According to physical and chemical properties, the materials can be divided into the metallic materials, inorganic nonmetallic materials, organic polymer materials, composite materials, etc.

According to application, the materials can be divided into the rail traffic materials, electronic materials, aerospace materials, nuclear materials, building materials, energy materials, biological materials, etc.

Furthermore, the above categories can be further classified with the deepening of researches. For example, the most commonly used metallic materials for the industrial production can be divided into two categories such as the ferrous metals and nonferrous metals. In the further classification, the ferrous metals are divided into iron, chromium, manganese and their alloys, and the nonferrous metals are divided into the other metals except the ferrous metals and their alloys e.g. aluminum, copper, gold, silver and their alloys (shown in Figure 7.1).

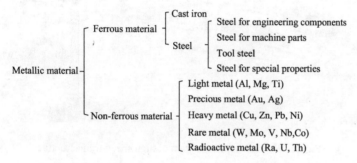

Figure 7.1 Classification of metallic material

(2) Introduction of Materials for rail traffic

Generally, the rail traffic refers to the vehicle running on the specific rails and operation system (shown in Figure 7.2). For the rail traffic, all kinds of parts such as vehicle body, wheels, axles, bogies, couplers, brake discs, pantographs, etc. are inextricably linked to the materials. The materials commonly used in rail traffic include the metallic materials, ceramic materials, composite materials, etc. The ceramic materials and composite materials will be described in more detail in Section 7.2. This section focuses on the related knowledge of metallic materials commonly used in rail traffic.

Figure 7.2 High speed train

More and more metallic materials are being applied in rail traffic. Let's take the body of vehicle as an example to explain as follows[1]. The vehicle body, which is the main part used as a place for accommodating the passengers and drivers and a foundation for installing and connecting the other equipment and components, consists of the underframe, two side walls (including doors and windows), roof, front and rear walls, etc. (shown in Figure 7.3) At present, the metallic materials used for the vehicle body mainly include stainless steel, high-strength weathering steel, low-alloy low-carbon high-strength steel, aluminum alloy, etc. The large hollow thin-walled aluminum materials, carbon fiber and composite materials are expected to be the host materials for vehicle body in the future. The lightening, safety and reliability of vehicle body and parts are the prerequisites for the safe operation of rail traffic vehicle at high speed and heavy load.

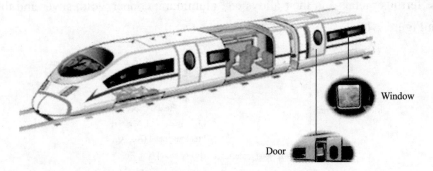

Figure 7.3   Schematic diagram of high speed train structure

7.1.2 Metallic Material

Metllic material

The main material for manufacturing machineries and key parts is the metallic material, which is known as engineering material with metallic properties and made by melting or sintering from metals or the metals added with non-metals. The commonly used metallic materials are iron carbon alloys (i.e. cast irons, steels, etc.), aluminum alloys, etc. especially in rail traffic.

(1) Properties of metallic material

The main properties of metallic material include mechanical property, physical property, chemical property, processing property, etc.

1) Mechanical property

The mechanical properties, referring to the properties of material under load, are ascertained by performing carefully designed laboratory experiments that replicate as nearly as possible the service conditions. These properties especially for the rail traffic vehicle mainly include the mechanical properties under static load, mechanical properties under dynamic load, mechanical property under alternate load, etc.

a) Mechanical property under static load

The mechanical properties under static load mainly include strength (such as yield strength, tensile strength, etc.), ductility (such as elongation, percent reduction in area, etc.), hardness (such as Rockwell hardness, Brinell hardness, microhardness, etc.), etc.

b) Mechanical property under dynamic load

The mechanical properties under dynamic load include impact toughness, impact energy, etc.

c) Mechanical property under alternate load

The mechanical property under alternate load mainly includes fatigue strength. Fatigue is a form of failure that occurs in structures subjected to dynamic and fluctuating stresses (i.e. bridges, aircraft and machine components). Under these circumstances it is possible for failure to occur at a stress level considerably lower than the tensile or yield strength for a static load. The term "fatigue" is used because this type of failure normally occurs after a lengthy period of repeated stress or strain cycling. Fatigue is important inasmuch as[2] it is the single largest cause of failure in metals, estimated to comprise approximately 90% of all metallic failures. Polymers and ceramics (except for glasses) are also susceptible to this type of failure. Furthermore, fatigue is catastrophic and insidious, occurring very suddenly and without warning.

The above mechanical properties measure the reliability and stability of materials under stress conditions from different aspects. The higher the mechanical properties of rail traffic materials are, the safer the vehicle becomes.

2) Physical property

The physical property mainly includes the thermal expansion, thermal conductivity, conductivity, etc. The requirements to physical properties of material may differ according to the service requirements of material.

a) Thermal expansion

Most solid materials expand upon heating and contract when cooled. The change in length with temperature for a solid material may be expressed as follows:

$$\frac{l_f - l_0}{l_0} = \alpha_l (T_f - T_0) \tag{7.1}$$

or

$$\frac{\Delta l}{l_0} = \alpha_l \Delta T \tag{7.2}$$

where l_0 and l_f represent, respectively, initial and final lengths with the temperature change from T_0 to T_f. The parameter α_l is called the linear coefficient of thermal expansion. It is a material property that is indicative of the extent to which a material expands upon heating, and has units of reciprocal temperature (°C^{-1}). Of course, heating or cooling affects all the dimensions of a body, with a resultant change in volume. Volume changes with temperature may be computed from

$$\frac{\Delta V}{V_0} = \alpha_v \Delta T \tag{7.3}$$

where ΔV and V_0 are the volume change and the original volume, respectively, and α_v symbolizes the volume coefficient of thermal expansion. In many materials, the value of α_v is anisotropic, that is, it depends on the crystallographic direction along which it is measured. For materials in which the thermal expansion is isotropic, α_v is approximately $3\alpha_l$.

From an atomic perspective, thermal expansion is reflected by an increase in the average distance between the atoms. This phenomenon can best be understood by consultation of the potential energy-versus-interatomic spacing curve for a solid material, and reproduced in Figure 7.4(a). The curve is in the form of a potential energy trough, and the equilibrium interatomic spacing at 0 K, r_0, corresponds to the trough minimum. Heating to successively higher temperatures (T_1, T_2, T_3, etc.) raises the vibrational energy from E_1 to E_2 to E_3, and so on. The average vibrational amplitude of an atom corresponds to the trough width at each temperature, and the average interatomic distance is represented by the mean position, which increases with temperature from r_0 to r_1 to r_2 and so on.

Thermal expansion is really due to the asymmetric curvature of this potential energy trough, rather than the increased atomic vibrational amplitudes with rising temperature. If the potential energy curve were symmetric [Figure 7.4(b)], there would be no net change in interatomic separation and, consequently, no thermal expansion.

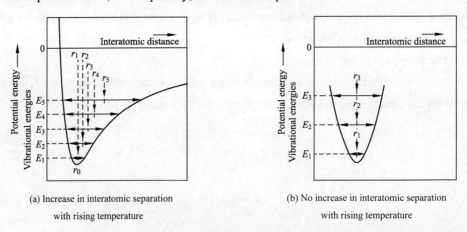

(a) Increase in interatomic separation with rising temperature

(b) No increase in interatomic separation with rising temperature

Figure 7.4　Potential energy-versus-interatomic spacing curve

For each class of materials (metals, ceramics, and polymers), the greater the atomic bonding energy is, the deeper and more narrow this potential energy trough is. As a result, the increase in interatomic separation with a given rise in temperature will be lower, yielding a smaller value of α_l. Table 7.1 lists the linear coefficients of thermal expansion for several materials. With regard to temperature dependence, the magnitude of the coefficient of expansion increases with rising temperature. The values in Table 7.1 are taken at room temperature unless indicated otherwise[3].

Table 7.1 Tabulation of the thermal properties for a variety of materials

Material	c_p (J/kg-K)[a]	α_l [(°C)$^{-1}$×10^{-6}][b]	k (W/m-K)[c]	L [Ω-W/(K)^2×10^{-8}]
Aluminum	900	23.6	247	2.20
Copper	386	17.0	398	2.25
Gold	128	14.2	315	2.50
Iron	448	11.8	80	2.71
Nickel	443	13.3	90	2.08
Silver	235	19.7	428	2.13
Tungsten	138	4.5	178	3.20
1025 Steel	486	12.0	51.9	—
316 Stainless steel	502	16.0	15.9	—
Brass (70Cu-30Zn)	375	20.0	120	—
Kovar (54Fe-29Ni-17Co)	460	5.1	17	2.80
Invar (64Fe-36Ni)	500	1.6	10	2.75
Super Invar (63Fe-32Ni-5Co)	500	0.72	10	2.68
Ceramics				
Alumina (Al$_2$O$_3$)	775	7.6	39	—
Magnesia (MgO)	940	13.5[d]	37.7	—
Spinel (MgAl$_2$O$_4$)	790	7.6[d]	15.0[e]	—
Fused silica (SiO$_2$)	740	0.4	1.4	—
Soda-line glass	840	9.0	1.7	—
Borosilicate (Pyrex^{TM}) glass	850	3.3	1.4	—
Polymers				
Polyethylene (high density)	1,850	106~198	0.46~0.50	—
Polypropylene	1,925	145~180	0.12	—
Polystyrene	1,170	90~150	0.13	—
Polytetrafluoroethylene (Teflon^{TM})	1,050	126~216	0.25	—
Phenol-formaldehyde, phenolic	1,590~1,760	122	0.15	—
Nylon 6.6	1,670	144	0.24	—
Polyisoprene	—	220	0.14	—

Note: [a] To convert to cal/g-K, multiply by 2.39×10^{-4}; to convert to Btu/lb$_m$-°F, multiply by 2.39×10^{-4}.

[b] To convert to (°F)$^{-1}$, multiply by 0.56.

[c] To convert to cal/s-cm-K, multiply by 2.39×10^{-3}; to convert to Btu/ft-h-°F, multiply by 0.578.

[d] Value measured at 100 °C.

[e] Mean value taken over the temperature range 0~1,000 °C.

b) Thermal conductivity

Thermal conduction is the phenomenon by which heat is transported from high- to low-temperature regions of a substance. The property that characterizes the ability of a material to transfer heat is the thermal conductivity. It is best defined in terms of the expression

$$q = -k\frac{dT}{dx} \tag{7.4}$$

where q denotes the heat flux, or heat flow, per unit time per unit area (area being taken as that perpendicular to the flow direction), k is the thermal conductivity, and dT/dx is the temperature gradient through the conducting medium.

The units of q and k are W/m^2 and $W/m \cdot K$, respectively. Equation 7.3 is valid only for steady-state heat flow—that is, for situations in which the heat flux does not change with time. Also, the minus sign in the expression indicates that the direction of heat flow is from hot to cold, or down the temperature gradient.

c) Conductivity

Conductivity i.e. electrical conductivity σ, is used to specify the electrical character of a material sometimes. It is simply the reciprocal of the resistivity, or

$$\sigma = \frac{1}{p} \tag{7.5}$$

and is indicative of the ease with which a material is capable of conducting an electric current (shown in Figure 7.5).

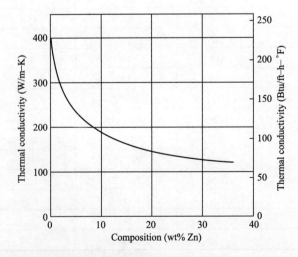

Figure 7.5　Thermal conductivity versus composition for copper-zinc alloys

Solid materials exhibit an amazing range of electrical conductivities, extending over 27 orders of magnitude, probably no other physical property experiences this breadth of variation. In fact, one way of classifying solid materials is according to the ease with which they conduct an electric current; within this classification scheme there are three groupings-conductors, semiconductors, and insulators. Metals are good conductors, typically having conductivities on the order of 10^7 $(\Omega \cdot m)^{-1}$. At the other extreme are materials with very low conductivities, ranging between 10^{-10} and 10^{-20} $(\Omega \cdot m)^{-1}$. These are electrical insulators. Materials with intermediate conductivities, generally from 10^{-6} to 10^4 $(\Omega \cdot m)^{-1}$, are termed semiconductors.

Room-temperature conductivities for several of the more common metals are contained in Table 7.2. Metals have high conductivities because of the large numbers of free electrons that have been excited into empty states above the Fermi energy.

Table 7.2 Room-temperature electrical conductivities for nine common metals and alloys

Metal	Electrical conductivity $[(\Omega \cdot m)^{-1}]$
Silver	6.8×10^7
Copper	6.0×10^7
Gold	4.3×10^7
Aluminum	3.8×10^7
Brass (70Cu-30Zn)	1.6×10^7
Iron	1.0×10^7
Platinum	0.94×10^7
Plain carbon steel	0.6×10^7
Stainless steel	0.2×10^7

At this point it is convenient to discuss conduction in metals in terms of the resistivity, the reciprocal of conductivity; the reason for this switch in topic should become apparent in the ensuing discussion.

Since crystalline defects serve as scattering centers for conduction electrons in metals, increasing their number raises the resistivity (or lowers the conductivity). The concentration of these imperfections depends on temperature, composition, and the degree of cold work of a metal specimen. In fact, it has been observed experimentally that the total resistivity of a metal is the sum of the contributions from thermal vibrations, impurities, and plastic deformation, that is, the scattering mechanisms act independently of one another. This may be represented in mathematical form as follows:

$$\rho = \rho_t + \rho_i + \rho_d \tag{7.6}$$

where ρ_t, ρ_i and ρ_d represent the individual thermal, impurity, and deformation resistivity contributions, respectively. Equation 7.6 is sometimes known as Matthiessen's rule. The influence of each variable on the total resistivity is demonstrated in Figure 7.6, a plot of resistivity versus temperature for copper and several copper-nickel alloys in annealed and deformed states. The additive nature of the individual resistivity contributions is demonstrated at -100 ℃.

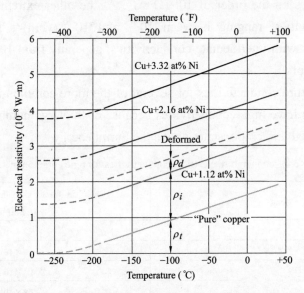

Figure 7.6　Electrical resistivity versus temperature for copper and three copper-nickel alloys

The above physical properties measure the functions of materials under applications from different aspects. The higher the physical properties of rail traffic materials are, the higher the grade of vehicle is.

3) Chemical property

The chemical property mainly refers to the corrosion resistance of material in various medium at normal temperature or high temperature such as corrosion resistance, oxidation resistance, etc.

a) Corrosion

Corrosion is defined as the destructive and unintentional attack of a metal. It is electrochemical and ordinarily begins at the surface. The problem of metallic corrosion is one of significant proportions. In economic terms, it has been estimated that approximately 5% of an industrialized nation's income is spent on corrosion prevention and the maintenance or replacement of products lost or contaminated as a result of corrosion reactions. The consequences of corrosion are all too common. Familiar examples include the rusting of vehicle body panels and radiator and exhaust components. As for the corrosion of materials, the lower the corrosion rate is, the better the corrosion resistance is.

b) Oxidation

Oxidation of metal alloys is also possible in gaseous atmospheres, normally air, wherein an oxide layer or scale forms on the surface of the metal. This phenomenon is frequently termed scaling, tarnishing, or dry corrosion.

4) Processing property

The processing property, which is a comprehensive reflection of physical, chemical and mechanical properties of metallic material during processing, refers to the performance suitable for cold working or hot working. It mainly includes casting property, forgability, weldability, machinability, etc. according to the process methods. In designing parts and selecting process methods, the processing properties of metallic material must be considered. For one example, grey cast iron has excellent casting property and hence it is widely used to make castings. However, it cannot be forged and welded for its poor forgability and weldability. For another example, low carbon steel has good weldability and hence it is widely used in welding structure while high carbon steel is not. The processing properties measure the manufacturing cost of materials from different aspects. The better the processing properties of rail traffic materials are, the faster the vehicle develops.

(2) Forming methods of metallic material

1) Common forming methods of metallic material

Casting, forging and welding are the most common forming methods for metallic materials.

a) Casting

Casting, also named as solidification molding, is a forming method for parts or roughcasts with certain shape, size and performance after the solidification of liquid metal in the mold. It is the oldest metal forming method in history and is still the main method of roughcast production nowadays.

Almost all metals are cast during some stage of the fabrication process. They can be cast either directly into the shape of the component or as ingots that can be subsequently shaped into a desired form. Casting offers several advantages over other methods of metal forming: It is adaptable to intricate shapes, to extremely large pieces, and to mass production; and it can provide parts with uniform physical and mechanical properties throughout. From an economic standpoint, it would be desirable to form most metal components directly from casting, since subsequent operations such as forming, extruding, annealing and joining add additional expense. However, since the growth rate of crystalline phases is much higher in the liquid state than in the solid state, the microstructure of as-cast materials is much coarser than that of heat-treated or annealed metals. Consequently, as-cast parts are usually weaker than wrought or rolled parts. Casting is generally only used as the primary fabrication process when ① the structural components will not be subjected to high stresses during service; ② high strength can be obtained by subsequent heat treatment; ③ the component is too large to be easily worked into shape; or ④ the shape of the component is too complicated to be easily worked into shape.

b) Forging

Forging is a forming method for forged pieces with certain shape, size and quality after the partial or total plastic deformation of billets and ingots under the action of pressurized equipment and tools or dies. This forming method can ensure the good mechanical properties of metal parts.

In all forging processes, the flow of metal is caused by application of an external force or pressure that pushes or pulls a piece of metal or alloy through a metal die. The pressure required to produce plastic flow is determined primarily by the yield stress of the material which, in turn, controls the load capacity of the machinery required to accomplish this desired change in shape. The pressure, P, used to overcome the yield stress and cause plastic deformation is given by

$$P=\sigma_y \varepsilon \tag{7.7}$$

where σ_y is the yield stress of the material and ε is the strain during the deformation. It is often useful to utilize the true strain in this relationship rather than the engineering strain. The pressure calculated in this way is generally much lower than that required in the actual process due to losses caused by friction, work hardening, and inhomogeneous plastic deformation. For example, the pressure predicted in Equation 7.7 is only 30%~55% of the working pressure required for extrusion, 50%~70% of that required for drawing, and 80%~90% of the pressure required for rolling.

In forging, the use of compressive forces means that tensile necking and fracture are generally avoided. Forging can be used to produce objects of irregular shape. There are two main categories of forging: hammering and pressing.

c) Welding

Welding is a connection method via atomic binding between work pieces by heating or pressing (or both) and plays a very important role in modern industrial production. For example, the manufacture of industrial products such as rail vehicle body, ship hull, building structure, etc. is inseparable from welding. In addition, welding is a permanent connection process which is irreplaceable by other methods.

In a sense, welding may be considered to be a fabrication technique. In welding, two or more metal parts are joined to form a single piece when one-part fabrication is expensive or inconvenient. Both similar and dissimilar metals may be welded. The joining bond is metallurgical (involving some diffusion) rather than just mechanical, as with riveting and bolting. A variety of welding methods exist, including arc and gas welding, as well as brazing and soldering.

During arc and gas welding, the workpieces to be joined and the filler material (i.e., welding rod) are heated to a sufficiently high temperature to cause both to melt; upon solidification, the filler material forms a fusion joint between the workpieces. Thus, there is a region adjacent to the weld that may have experienced microstructural and property alterations; this region is termed the heat-affected zone (HAZ). Possible alterations include:

- If the workpiece material was previously cold worked, this heat-affected zone may have experienced recrystallization and grain growth, and thus a diminishment of strength, hardness, and toughness. The HAZ for this situation is represented schematically in Figure 7.7.
- Upon cooling, residual stresses may form in this region that weaken the joint.
- For steels, the material in this zone may have been heated to temperatures sufficiently high so as to form austenite. Upon cooling to room temperature, the microstructural products that form depend on cooling rate and alloy composition. For plain carbon steels, normally pearlite and a proeutectoid phase will be present. However, for alloy steels, one microstructural product may be martensite, which is ordinarily undesirable because it is so brittle.
- Some stainless steels may be "sensitized" during welding, which renders them susceptible to intergranular corrosion.

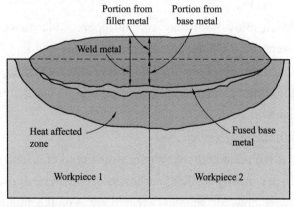

Figure 7.7　Schematic cross-sectional representation showing the zones in the vicinity of a typical fusion weld

2) Classification according to forming temperature

Metal forming can be divided into cold working and hot working according to the forming temperature.

a) Recrystallization

When the metal is reheated to a certain temperature after plastic deformation, the elongated grains will become equiaxed grains with the same lattice structure before deformation by re-nucleation and crystallization. The minimum temperature, at which a metal with a cold deformation of more than 70% can completely recrystallized by 1 hour heating, is usually defined as the recrystallization temperature. The experimental results show that the recrystallization temperature of metal T_m can be approximately calculated by $T_m = 0.3 \sim 0.4T$ (melting point).

b) Cold working

Cold working is a forming manner in which the forming temperature is below the recrystallization temperature of metal. It includes cold rolling, cold forging, stamping, cold extrusion, turning, milling, grinding, numerical control machining, etc.

Turning

c) Hot working

Hot working is a forming manner in which the forming temperature is higher than the recrystallization temperature of metal. It includes casting, hot forging, hot rolling, welding, metal heat treatment, hot cutting, hot extrusion, etc.

(3) Industrial steel

From the production of metallic material, the production of steel (carbon content ≈ 0.02 ~ 2.11 wt%) and iron (carbon content > 2.11 wt%) is absolutely dominant, accounting for about 95% of the total production of metal. Owing to the low price relative to other metal alloys and the excellent comprehensive performance which can meet the most requirements of engineering application, the industrial steel has been widely applied.

Grinding

1) Classification of steel

Steels are ironcarbon alloys that may contain appreciable concentrations of alloying elements. According to chemical composition, steel can be divided into carbon steel and alloy steel. According to application, steel can be divided into structural steel, tool steel and special performance steel. According to process manner, steel can be divided into cast steel, forged steel, rolled steel, cold drawn steel, etc.

2) Common steel for railway vehicle body

The vehicle body is the main body of vehicle structure. At present, the bodies of railway vehicle under 200 km/h are mainly made of carbon steel in China. Those of 200 to 250 km/h are generally carbon steel body or stainless steel body. For the high-speed multiple units above 250km/h, stainless steel bodies or aluminum alloy bodies are generally adopted.

a) Carbon steel vehicle body

The main materials used for carbon steel vehicle body are ordinary carbon steel and nickel-chromium series low-alloy weathering steel.

The weathering steel i.e. atmospheric corrosion resistant steel, which belongs to low alloy steel between ordinary steel and stainless steel, results from the ordinary carbon steel adding a small amount of corrosion resistance element such as chromium, nickel, etc. Besides the features of high quality steel such as good toughness, plasticity, formability, weldability, wear resistance, high temperature resistance, fatigue resistance, etc., it has excellent atmospheric corrosion resistance which is about 2~5 times that of ordinary carbon steel. Furthermore, the longer it serves, the more prominent its atmospheric corrosion resistance displays. Therefore, the weathering steel can be applied to the steel structures such as vehicles, bridges, towers, etc. which are exposed to the atmosphere for a long time.

The composition of weathering steel is list in Table 7.3.

Table 7.3 Composition of weathering steel (wt%)

C	Si	Mn	S	P	Cu	Cr	Ni	Fe
≤0.12	0.25~0.75	0.2~0.5	≤0.02	0.06~0.12	0.25~0.5	0.3~1.25	0.12~0.65	balance

The strength and stiffness of vehicle body are related to the safety and reliability of vehicle. Due to its high strength and elastic modulus, steel can guarantee the safety and reliability of rail traffic vehicle. However, the carbon steel vehicle body, whose weight and welding deformation are not too ideal, has a limitation on application and development.

b) Stainless steel vehicle body

Stainless steel, possessing excellent values in corrosion resistance, heat resistance, stress resistance, weldability, rolling formability, drawability, etc. is gradually replacing carbon steel to produce vehicle body, especially in the moist, weak acid and high temperature areas. However, the corrosion resistance of stainless steel decreases with the increase of carbon content. In addition, the main alloy elements such as Cr and Ni, etc. can play a role in corrosion resistance, strength, hardness and plasticity only when their contents reach a certain value. Therefore only two kinds of stainless steel are used in vehicle body i.e. SUS301L and SUS304.

SUS301L and SUS304 are austenitic stainless steel whose composition and mechanical properties are shown in Table 7.4 and Table 7.5. Austenitic stainless steel refers to the stainless steel with austenite structure at room temperature. The stainless steel, which contains 18 wt% chromium and 8 wt% nickel, is the most typical austenitic stainless steel and is often called 18-8 stainless steel. This kind of stainless steel has very high toughness and ductility in a wide range of temperature and thus can be easily cold rolled and compressed. Its corrosion resistance and acid resistance are better than those of ferritic stainless steel and martensitic stainless steel. Especially, the Ni-Cr series austenitic stainless steel will have better corrosion resistance and acid resistance after adding molybdenum and copper. Besides the Ni-Cr series austenite stainless steels, there are Cr-Ni-Mn series and Fe-Al-Mn series austenite stainless steels. Austenitic stainless steel has been widely used in industry because of its good comprehensive properties.

Table 7.4　Composition of stainless steel (wt%)

Steel	C	Si	Mn	S	P	N	Cr	Ni	Fe
SUS301L	≤0.03	≤1	≤2	≤0.03	≤0.015	≤0.2	16~18	6~8	balance
SUS304	≤0.08	≤1	≤2	≤0.03	≤0.015		18~20	6~10.5	balance

Table 7.5　Mechanical properties of stainless steel

Steel	Modified treatment	Yield strength (N/mm^2)	Tensile strength (N/mm^2)	Elongation (%)		
				Thickness <0.4 mm	Thickness >0.4 mm <0.8 mm	Thickness >0.8 mm
SUS301L	1/1H	>345	>690		>40	
	1/2H	>410	>760		>35	
	3/1H	>480	>820		>25	
	H	>685	>930		>20	
SUS304		>205	>520		>40	

(4) Aluminum alloy

Aluminum alloy is one of the most widely used nonferrous metal structural materials in industrial production. It has been widely used in aviation, aerospace, rail traffic, machinery manufacturing, shipping and chemical industries.

Aluminum and its alloys are characterized by a relatively low density (2.7 g/cm^3 as compared to 7.9 g/cm^3 for steel), high electrical and thermal conductivities, and a resistance to corrosion in some common environments, including the ambient atmosphere. Many of these alloys are easily formed by virtue of high ductility; this is evidenced by the thin aluminum foil, into which the relatively pure material may be rolled. Since aluminum has an FCC (faced centered cubic) crystal structure, its ductility is retained even at very low temperatures. The chief limitation of aluminum is its low melting temperature 660℃, which restricts the maximum temperature at which it can be used.

1) Classification of aluminum and aluminum alloys

a) Classification of pure aluminum

Classification of pure aluminum is shown in Table 7.6.

Table 7.6　Classification of pure aluminum

Name	Purity	Application
High-purity aluminum	99.93%~99.99%	Scientific research, electrical apparatus manufacture
Industrial high-purity aluminum	98.85%~99.9%	Aluminum foil, raw aluminum alloy
Industrial purity aluminum	98.0%~99.0%	Wires, cables

b) Classification of aluminum alloy

The mechanical strength of aluminum may be enhanced by cold work and by alloying; however, both processes tend to diminish resistance to corrosion. Principal alloying elements include copper, magnesium, silicon, manganese, and zinc. Some alloying elements increase its strength by solid-solution strengthening. Others increase its strength by precipitation hardening as a result of alloying. In several of these alloys precipitation hardening is due to the precipitation of two elements other than aluminum, to form an intermetallic compound such as $MgZn_2$.

According to the content of alloying elements and processing characteristics, aluminum alloy can be divided into wrought aluminum alloy and casting aluminum alloy (shown in Figure 7.8).

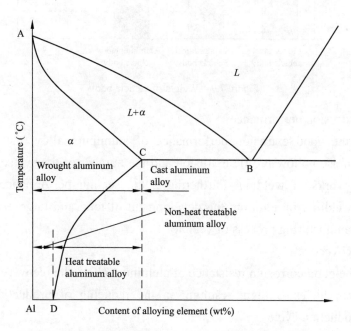

Figure 7.8 Phase diagram of aluminum alloy

2) Application of aluminum alloy in rail traffic

With the development of high speed railway technology, the manufacturing technology of high speed trains has been improved rapidly. The manufacturing materials of train have changed greatly. For example, the manufacturing material of train body has realized the transformation from carbon steel to stainless steel and aluminum alloy. In particular, aluminum alloy body is widely used for its light weight, excellent corrosion resistance and good appearance smoothness. At present, the proportion of aluminum alloy vehicle body in railway transportation field is gradually increasing, accounting for 90% in high speed train and 25% in urban rail train.

a) Advantages of aluminum alloy

(a) Light weight

Compared with steel vehicle body, the weight of aluminum alloy vehicle body can be reduced by about 30% (shown in Figure 7.9). The lightweight vehicle can save energy, reduce the braking force, improve the power performance of vehicle, debase noise, diminish pollution and protect the environment.

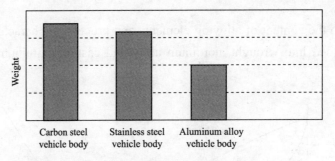

Figure 7.9 Weight of vehicle body

(b) Good extrusion performance

Owing to the good extrusion performance of aluminum alloy, the aluminum alloy vehicle body can use hollow profile extrusion and the number of transverse parts is reduced as well as the works of welding. Furthermore, the sealing performance of structure is increased after welding. In addition, the development of long and large extrusion part also simplifies the manufacturing process of vehicle.

(c) High recovery rate

Owing to the good corrosion resistance of aluminum alloy, the recovery rate of scrapped aluminum alloy vehicles is high, resulting in the reducing of production cost and the shortening of production cycle.

(d) Excellent comprehensive performance

By adding Mg, Zn and Mn (shown in Table 7.7), the comprehensive performance of aluminum alloys is improved greatly (shown in Table 7.8). The vehicle interior decoration pieces, inner wall plates, door plates, brackets, flaps, window frames, etc. begin to use aluminum alloy, resulting in the reduction of the vehicle weight and the system energy consumption.

Table 7.7 Composition of aluminum alloy (wt%)

Material	Si	Fe	Cu	Mn	Mg	Cr	Zn	Ti	Al
5052	<0.25	<0.40	<0.10	<0.10	2.2~2.8	0.15~0.35	<0.10		Balance
6061	0.4~0.8	<0.70	0.15~0.40	<0.15	0.8~1.2	0.04~0.35	<0.25	<0.15	Balance

(continued table)

Material	Si	Fe	Cu	Mn	Mg	Cr	Zn	Ti	Al
6063	0.2~0.6	<0.35	<0.10	<0.10	0.45~0.9	<0.10	<0.10	<0.10	Balance
7003	<0.30	<0.35	<0.20	<0.30	0.5~1.0	<0.20	5.0~6.5	<0.20	Balance
7N01	<0.30	<0.35	<0.20	0.20~0.70	1.05~2.0	<0.30	4.0~5.0	<0.20	Balance

Table 7.8 Mechanical properties of aluminum alloy

Material	Quality level	Thickness (mm)	Yield strength (N/mm^2)	Tensile strength (N/mm^2)	Ductility (%)
5052	II II2	-	>70	175	
	0	-	>70	175~245	>20
6061	0		<110	<145	>16
	T4		>110	>175	>16
	T42		>85	>175	>16
	T6		>245	>265	>8
	T62		>245	>265	>10
6063	T1	<12	>60	>120	>12
		12~25	>55	>110	>12
	T5	<12	>110	>115	>8
		12~25	>110	>145	>8
	T6	<3	>117	>205	>8
		3~25	>175	>205	>10
7003	T5	<12	>245	>285	>10
		12~25	>235	>275	>10
7N01	0		>145	<245	>12
	T4		>195	>315	>11
	T5		>245	>325	>10
	T6		>275	>335	>10

b) Application of aluminum alloy in rail traffic

The aluminum alloy vehicle body is being gradually applied in high-speed railway. After the first Shinkansen in Paris-Lyon, France builds the second Shinkansen i.e. Atlantic Shinkansen in which aluminum alloy vehicle body puts into use in November 1989. The speed of this second generation TGV-Atlantique reaches 300 km/h, In May 1990, the TGV-Atlantique set a world record of 515.3 km/h which is kept for 17 years. On April 3, 2007, the French TGV-V150 set a new record of 574.8 km/h, resulting from the continuous application of aluminum alloy (shown in Figure 7.10).

Figure 7.10　French TGV-V150

The 500 Series is the most popular vehicle in Shinkansen of Japan (shown in Figure 7.11). It uses aluminum alloy body and is the fastest vehicle. Of all Shinkansen vehicles, the best streamlined one is the 500 Series. Its front streamlined bending part is more than 9 meters long. The aluminum alloy vehicle body of 700 Series uses the latest space truss structure in which there is no beam columns on the roof and side walls (shown in Figure 7.12). The aluminum profiles of this complex shaped aluminum alloy body are welded by friction stir welding.

Figure 7.11　500 Series vehicle

Figure 7.12　700 Series vehicle

In China, the aluminum alloy bodies have been used in both the CRH2 [shown in Figure 7.13(a)] EMUs running at a speed of 200 km/h produced by Sifang Co., Ltd. of South Car and the CRH5 [shown in Figure 7.13(b)] EMUs running at a speed of 250 km/h produced by Changchun Rail Co., Ltd.

Chapter 7 Materials Science and Technology
第7章 材料科学与工程

(a) CRH2

(b) CRH5

Figure 7.13 CRH

c) Comparison of indexes between stainless steel body and aluminum alloy body

The indexes of stainless steel and aluminum alloy are shown in Table 7.9.

Table 7.9 Indexes of stainless steel body and aluminum alloy body

Index name	Aluminum alloy	Stainless steel
Density	2.71 g/cm^3	7.85 g/cm^3
Tensile strength	270~310 MPa	690~930 MPa
Yield strength	200~260 MPa	345~685 MPa
Rigidity (elasticity modulus)	0.71E10 N/mm^2	2.06E10 N/mm^2
Melting point	About 660 ℃	About 1,500 ℃
Welding manner	Argon arc welding	Spot welding, laser welding
Corrosion resistance	Bad	Good
Price	About 130,000 per ton, cheap	About 170,000 per ton, expensive
Service life	Short	Long

With the development of modern material and production technology, there will be more and more optional materials for rail vehicle. Attention should be paid to a comprehensive consideration as selecting material of rail vehicle, that is, on the basis of safety, lightweight and environmental protection, the comfortability and beauty should be satisfied to the maximum extent.

Words and Expressions

structural material [ˈstrʌktʃərəl məˈtɪərɪəl]	结构材料
functional material [ˈfʌŋkʃənəl məˈtɪərɪəl]	功能材料
metallic material [məˈtælɪk məˈtɪərɪəl]	金属材料
inorganic nonmetallic material [ˌɪnɔːˈgænɪk ˈnɒnmɪˈtælɪk məˈtɪərɪəl]	无机非金属材料

composite [ˈkɒmpəzɪt] a conceptual whole made up of complicated and related parts	n. 复合材料
ferrous metal [ˈferəs ˈmetl]	黑色金属
nonferrous metal [ˈnɒnˈferəs ˈmetl]	有色金属
aluminum alloy [əˈluːmɪnəm ˈælɔɪ]	铝合金
vehicle body [ˈviːɪkl bɒdi]	车体
ceramic material [səˈræmɪk məˈtɪərɪəl]	陶瓷材料
stainless steel [ˌsteɪnləs ˈstiːl]	不锈钢
high-strength weathering steel [haɪ streŋθ ˈweðərɪŋ stiːl]	高强度耐候钢
low-alloy high-strength steel [ləʊ ˈælɔɪ haɪ streŋθ stiːl]	低合金高强度钢（简称为 HSLA steel）
austenite stainless steel [ˈɔːstə naɪt ˈsteɪnlɪs stiːl]	奥氏体不锈钢
engineering material [ˌendʒɪˈnɪərɪŋ məˈtɪərɪəl]	工程材料
iron carbon alloy [ˈaɪən ˈkɑːbən ˈælɔɪ]	铁碳合金
yield strength [jiːld streŋθ]	屈服强度
tensile strength [ˈtensəl streŋθ]	抗拉强度
fatigue strength [fəˈtiːg streŋθ]	疲劳强度
ductility [dʌkˈtɪləti] the malleability of something that can be drawn into threads or wires or hammered into thin sheets	n. 韧性
elongation [iːlɒŋˈgeɪʃən] the quality of being elongated	n. 延伸率
percent reduction in area [pəˈsent rɪˈdʌkʃən ɪn ˈeərɪə]	断面收缩率
Brinell hardness(HB) [ˈbrɪnel ˈhɑːdnɪs]	洛氏硬度
Rockwell hardness(HR) [ˈrɒkwel ˈhɑːdnɪs]	布氏硬度
micro hardness [ˈmaɪkrəʊ ˈhɑːdnɪs]	显微硬度
impact toughness [ˈɪmpækt ˈtʌfnəs]	冲击韧度
impact energy [ˈɪmpækt ˈenədʒi]	冲击功
weldability [weldə ˈbɪlɪti] of a material refers to its ability to be welded	焊接性能
machinability [məˌʃiːnəˈbɪlɪti] the ease with which a metal can be machined to an acceptable surface	切削加工性能

vacuum electron beam welding [ˈvækjuəmˈiˈlektrɔnˈbiːmˈweldɪŋ]	真空电子束焊接
recrystallization temperature [rekrɪstəlaɪˈzeɪʃən ˈtemprətʃə]	再结晶温度
carbon steel [ˈkɑːbən stiːl]	碳素钢
wrought aluminum alloy [rɔːt əˈluːmɪnəm ˈælɔɪ]	变形铝合金
casting aluminum alloy [ˈkæstɪŋ əˈluːmɪnəm ˈælɔɪ]	铸造铝合金
space truss structure [speɪs trʌs ˈstrʌktʃə]	网架结构

Notes

[1] 本句是祈使句，全句可译为：以车体为例说明如下。

[2] inasmuch as 可译为"由于"。

[3] unless indicated otherwise 可译为"除非另有说明"。

Questions for Discussion

1. Which kinds of materials are the commonly used metallic materials?
2. Which mechanical properties does the rail traffic vehicle mainly require?
3. Which kinds of forming methods are commonly used in the forming of metal materials?
4. How many crystallization zones are there in a solidified metal ingot?
5. What are the main advantages of casting?
6. How does the welding can be divided?
7. What are the main advantages of laser welding?
8. What is the hardest carbon steel?
9. What is the chief limitation in application of aluminum?
10. What is the reason for the wide application of aluminum alloys?

7.2 Ceramic Material and Composite Material

7.2.1 Ceramic Material

(1) Introduction

Ceramic material refers to a kind of inorganic nonmetallic material which is made from natural or synthetic compounds through processing or high-temperature sintering. Most ceramics are compounds between metallic and nonmetallic elements for which the interatomic bonds are either totally ionic, or predominantly ionic but having some covalent character. The term "ceramic" comes from the Greek word "keramikos", which means "burnt stuff", indicating that desirable properties of these materials are normally achieved through a high-temperature heat treatment process called firing.

(2) Properties of ceramic material

1) Mechanical property

Ceramic material is the engineering material with the highest hardness and the best stiffness. Most ceramic materials have a hardness of more than 1,500 HV (polymer hardness < 20 HV; quenched steel hardness = 500~800 HV) and the hardness of some ceramic materials can even reach up to 5,000 HV. All ceramic materials have the best stiffness, but they almost have no plasticity and ductility at room temperature. In addition, ceramic material has high compressive strength, large flexural strength and excellent temperature strength. Ceramic materials are somewhat limited in applicability by their mechanical properties, which in many respects are inferior to those of metals. The principal drawback is a disposition to catastrophic fracture in a brittle manner with very little energy absorption.

a) Brittle fracture of ceramic

At room temperature, both crystalline and noncrystalline ceramics almost always fracture before any plastic deformation can occur in response to an applied tensile load. The brittle fracture process consists of the formation and propagation of cracks through the cross section of material in a direction perpendicular to the applied load. Crack growth in crystalline ceramics may be either transgranular (i.e. through the grains) or intergranular (i.e. along grain boundaries); for transgranular fracture, cracks propagate along specific crystallographic (or cleavage) planes, planes of high atomic density.

b) Flexural strength

The stress-strain behavior of brittle ceramics is not usually ascertained by a tensile test for three reasons. First, it is difficult to prepare and test specimens having the required geometry. Second, it is difficult to grip brittle materials without fracturing them; and third, ceramics fail after only about 0.1% strain, which necessitates that tensile specimens be perfectly aligned to avoid the presence of bending stresses, which are not easily calculated. Therefore, a more suitable transverse bending test is most frequently employed, in which a rod specimen having either a circular or rectangular cross section is bent until fracture using a three- or four-point loading technique.

c) Elastic behavior

The elastic stress-strain behavior for ceramic materials using these flexure tests is similar to the tensile test results for metals: a linear relationship exists between stress and strain. Figure 7.14 compares the stress-strain behavior to fracture for aluminum oxide and glass. Again, the slope in the elastic region is the modulus of elasticity; the range of moduli of elasticity for ceramic materials is between about 70 and 500 GPa, being slightly higher than for metals. Also, from Figure 7.14 note that neither material experiences plastic deformation prior to fracture.

d) Influence of porosity

For some ceramic fabrication techniques, the precursor material is in the form of a powder. Subsequent to compaction or forming of these powder particles into the desired shape, pores or void spaces will exist between the powder particles. During the ensuing heat treatment, much of this porosity will be eliminated; however, it is often the case that this pore elimination process is incomplete and some residual porosity will remain.[1]

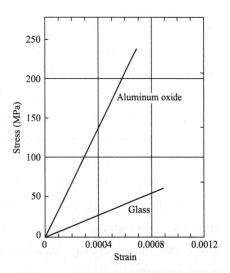

Figure 7.14 Typical stress-strain behavior to fracture for aluminum oxide and glass

e) Hardness

One beneficial mechanical property of ceramics is their hardness, which is often utilized when an abrasive or grinding action is required; in fact, the hardest known materials are ceramics. Only ceramics having Knoop hardnesses of about 1,000 or greater are utilized for their abrasive characteristics (shown in Table 7.10).

Table 7.10 Approximate knoop hardness (100 g load) for seven ceramic materials

Material	Approximate Knoop hardness
Diamond (carbon)	7,000
Boron carbide (B_4C)	2,800
Silicon carbide (SiC)	2,500
Tungsten carbide (WC)	2,100
Aluminum oxide (Al_2O_3)	2,100
Quartz (SiO_2)	800
Glass	550

2) Thermal performance

Ceramic material generally has a high melting point of more than 2000 ℃ and has excellent chemical stability at high temperature. Relatively strong interatomic bonding forces are found in many ceramic materials as reflected in comparatively low coefficients of thermal expansion; values typically range between about 0.5×10^{-6} and 15×10^{-6} (℃). For noncrystalline ceramics and also those having cubic crystal structures, is isotropic. Otherwise, it is anisotropic; and, in fact, some ceramic materials, upon heating, contract in some crystallographic directions while expanding in others. For inorganic glasses, the coefficient of expansion is dependent on composition. Fused silica (high-purity SiO_2 glass) has a small expansion coefficient, 0.4×10^{-6} (℃). This is explained by a low atomic packing density such that interatomic expansion produces relatively small macroscopic dimensional changes. Ceramic materials that are to be subjected to temperature changes must have coefficients of thermal expansion that are relatively low, and in addition, isotropic. Otherwise, these brittle materials may experience fracture as a consequence of nonuniform dimensional changes in what is termed thermal shock.

The thermal conductivity of ceramic is lower than that of metal and thus ceramic is a good thermal insulation material. Furthermore, the linear expansion coefficient of ceramic is lower than that of metal. Accordingly, the ceramic parts have good dimensional stability under operating temperature variation.

3) Electrical characteristics

Most ceramics have good electrical insulation, so they are widely used to fabricate various insulation devices used at 1~110 kV. Both cations and anions in ceramic materials possess an electric charge and, as a consequence[2], are capable of migration or diffusion when an electric field is present. Thus an electric current will result from the net movement of these charged ions, which will be present in addition to current due to any electron motion. Of course, anion and cation migrations will be in opposite directions. The total conductivity of an ionic material σ_{total} is thus equal to the sum of both electronic and ionic contributions, as follows:

$$\sigma_{total} = \sigma_{electronic} + \sigma_{ionic} \tag{7.8}$$

Either contribution may predominate depending on the material, its purity, and, of course, temperature.

Thus, the ionic contribution to the total conductivity increases with increasing temperature, as does the electronic component. However, in spite of the two conductivity contributions, most ionic materials remain insulative, even at elevated temperatures.

Ferroelectric ceramic ($BaTiO_3$) have a high dielectric constant and can be used to make capacitors. In addition, ferroelectric ceramic has piezoelectric property, can change its shape under the action of external electric field and can convert the electrical energy into the mechanical energy. They can be used for amplifier, record player, ultrasonic instrument, sonar, medical sonograph, etc. Some ceramics such as NiO, Fe_3O_4, etc. also have the characteristics of semiconductor and can be used for rectifier and semiconductor.

4) Chemical and physical properties

Ceramic material has excellent oxidation resistance and cannot oxidize at high temperature. Ceramic materials, being compounds between metallic and nonmetallic elements, may be thought of as having already been corroded. Thus, they are exceedingly immune to corrosion by almost all environments, especially at room temperature. Corrosion of ceramic materials generally involves simple chemical dissolution, in contrast to the electrochemical processes found in metals. Furthermore, it has good corrosion resistance to acid, alkali and salt. Ceramic materials are frequently utilized because of their resistance to corrosion. Glass is often used to contain liquids for this reason. Refractory ceramics must not only withstand high temperatures and provide thermal insulation but, in many instances, must also resist high-temperature attack by molten metals, salts, slags, and glasses[3]. Some of the new technology schemes for converting energy from one form to another that is more useful require relatively high temperatures, corrosive atmospheres, and pressures above the ambient. Ceramic materials are much better suited to withstand most of these environments for reasonable time periods than are metals.

Some ceramic materials such as $MgFe_2O_4$, $CuFe_2O_4$, Fe_3O_4, etc. have excellent magnetic property and have been used for recording tape, record, transformer core, large computer memory device, etc.

5) Optical properties

By virtue of their electron energy band structures, ceramic materials may be transparent to visible light. Some ceramic materials have excellent optical conductivity. They have been used for solid-state laser, optic fiber, optical storage, high pressure sodium lamp tube, etc.

(3) Ceramic forming process

Since ceramic materials have relatively high melting temperatures, casting them is normally impractical. Furthermore, in most instances the brittleness of these materials precludes deformation. The most commonly used processing technology is powder metallurgy.

Powder metallurgy is the process of converting powders into ingots or finished components via compaction and sintering. Powder metallurgy is used whenever porous parts are needed, whenever the parts have intricate shapes, whenever the mixture cannot be achieved in any other manner, or whenever the mixtures have very high melting points[4]. The technique is suitable to virtually any material, however. An organic binder is frequently added to the metallic powders during the mixing state to facilitate compaction, but volatilizes at low temperatures during the sintering process. The advantages of powder metallurgy included improved microstructures and improved production economies.

(4) Classification of ceramic

Ceramic is the general name of pottery and porcelain. In the traditional concept, ceramic refers to all the handiworks from the coarsest earth ware to the finest porcelain which are made from clay or other inorganic nonmetallic materials.

Up to now, ceramic can be described as follows: Ceramic is an artificial rock with a certain pattern, which is made of aluminosilicates or oxides by a specific physical and chemical process at some temperature and atmosphere according to human intentions.

1) Classification according to application

Ceramic can be divided into structural ceramic, functional ceramic etc.

a) Structural ceramic

Structural ceramic refers to the advanced ceramic with excellent mechanical, thermal and chemical properties such as high temperature resistance, erosion resistance, corrosion resistance, high hardness, high strength, low creep rate, etc., which is commonly used in various structural components. For example, the structural ceramic using alumina as the raw material has high strength and hardness, good oxidation resistance, corrosion resistance, wear resistance and ablation resistance especially at high temperature. They can withstand a high temperature of 1,980 ℃ in air and are the important material in the fields of space, military, atomic energy and chemical industry. The structural ceramic has attracted a lot of attention in the material industry and its application is getting wider and wider.

Nowadays, many kinds of structural ceramics have been developed. After long application in practice, silicon nitride ceramic is considered as the promising structural ceramics.

b) Functional ceramic

Functional ceramic refers to the ceramic material with special properties in electricity, magnetism, light, heat, chemistry, biology, etc., which is commonly used in various functional components. With the rapid development of materials science, various new properties and new applications of functional ceramic materials have been recognized and actively developed.

Many kinds of functional ceramics with special properties have been used in different fields. Here is a brief description as follows.

The conductive ceramics, semiconductor ceramics, dielectric ceramics and insulating ceramics have been used to make the various electronic parts such as capacitors, resistors, high-temperature and high-frequency components, voltage transformers, etc.

Piezoelectric ceramic is a kind of functional ceramic that can convert pressure into electricity. Even the small pressure produced by acoustic vibration can realize the deformation of piezoelectric ceramic on the surface of which an electricity can be created. A gas electronic lighter can fire thousands of times as the flint is substituted with the piezoelectric ceramic.

Transparent ceramic is a kind of functional ceramic whose main composition includes magnesium oxide, calcium oxide, calcium fluoride, etc. The transparent ceramic with a very high mechanical strength and hardness can pass through light and be used to produce transparent bullet-proof parts. In addition, it can also be used to manufacture high-speed cutting tools, jet engine parts, tank observation windows, etc. and can even replace stainless steel.

2) Classification according to the source of raw material

According to the source of raw materials, ceramic can be divided into ordinary ceramic and special ceramic.

Ordinary ceramic is the ceramic that uses natural aluminosilicate minerals (clay, feldspar, quartz, etc.) as raw materials. Therefore, it is also known as aluminosilicate ceramic.

Special ceramic is an inorganic material with fine crystalline structure, which uses synthetic compounds with high purity (such as Al_2O_3, ZrO_2, SiC, Si_3N_4, BN, etc.) as raw materials. It has a series of superior physical, chemical and biological properties. Its application is much wider than that of ordinary ceramic. This kind of ceramic is also called special ceramic or fine ceramic.

(5) New ceramic materials

1) Special features of new ceramic material

In addition to the versatile properties of ceramic, new ceramic also has some unique performance as follows:

- It has special heat and mechanical properties such as high-temperature resistance and heat-insulating property, excellent hardness and wear resistance, etc.
- It has special electrical properties such as insulativity, piezoelectricity, semiconductive property.
- It has special chemical functions such as catalysis, anti-corrosion, adsorption, etc.
- It has special biological properties such as biocompatibility and can be used as biological structural materials.

2) Classification of new ceramic material

According to chemical composition, the new ceramic material can be divided into pure oxide ceramic material and non-oxide ceramic material.

According to the performance and feature, the new ceramic material can be divided into super hard ceramic material, high toughness ceramic material, semiconductor ceramic material, electrolyte ceramic material, magnetic ceramic material, conductive ceramic material, etc.

3) Development of new ceramic material

With the development of science and technology, new ceramic material enters the nanometer age. Nanometer (10^{-9} m) is a measure of scale. Nanomaterial is a kind of material in which the size of grains is less than 100 nm at least for one phase. With the development of new ceramic materials, nanoceramic materials such as nano calcium fluoride and nano zinc oxide have been developed. Nanoceramic material refers to the ceramic material whose average size of grains is less than 100 nm and has become an important branch of material.

Besides the excellent properties of conventional ceramic materials, nanoceramic materials also have high toughness. It is possible for nanoceramic materials to realize the processing such as forging, extrusion, drawing, bending, etc. just like that of metals. Thus nanoceramic materials can be used to directly produce the parts without machining. This will completely eliminate the "fragile" figure for ceramic.

The sintering property of nanoceramic material is excellent and its sintering temperature is decreased by several hundred degrees Kelvin compared with ordinary ceramic material. Many nanoceramics can be sintered below 1,273 K. Low-temperature sintering can save energy, purify environment and even can be used to realize the strong connection of different ceramic materials.

In short, new ceramic materials are being widely used in many fields and will have a wider application range with the development of modern science and technology.

7.2.2 Composite

(1) Definition of composite

Composite

Composite is a multi-phase material with the comprehensive property resulted from the combination of two or more material components with different physical and chemical properties by some advanced material processing technology. The composite consists of matrix phase and reinforcing phases. The component of composite includes metal, polymer or ceramic.

1) Composite constituents

The constituents in a composite retain their identity such that they can be physically identified and they exhibit an interface between one another. This concept is graphically summarized in Figure 7.15. The body constituent gives the composite its bulk form, and it is called the matrix. The other component is a structural constituent, sometimes called the reinforcement, which determines the internal structure of the composite. Though the structural component in Figure 7.15 is a fiber, there are other geometries that the structural component can take on. The region between the body and structural constituents is called the interphase. It is quite common (even in the technical literature), but incorrect, to use the term interface to describe this region. An interface is a two-dimensional

Composite materials

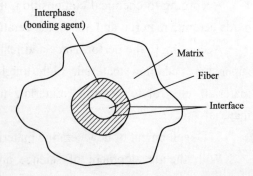

Figure 7.15 Schematic illustration of composite constituents

construction—an area having a common boundary between the constituents—whereas an interphase is a three-dimensional phase between the constituents and, as such, has its own properties. It turns out that these interphase properties play a very important role in determining the ultimate properties of the bulk composite. For instance, the interphase is where mechanical stresses are transferred between the matrix and the reinforcement. The interphase is also critical to the long-term stability of a composite. It will be assumed that there is always an interphase present in a composite, even though it may have a thickness of only an atomic dimension.

The chemical composition of the composite constituents and the interphase is not limited to any particular material class. There are metal-matrix, ceramic-matrix, and polymer-matrix composites, all of which find industrially relevant applications. Similarly, reinforcements in important commercial composites are made of such materials as steel, E-glass, and Kevlar. Many times a bonding agent is added to the fibers prior to compounding to create an interphase of a specified chemistry.

2) Combination effects in composite

There are three ways that a composite can offer improved properties over the individual components, collectively called combination effects. A summation effect arises when the contribution of each constituent is independent of the others. For example, the density of a composite is, to a first approximation, simply the weighted average of the densities of its constituents. The density of each component is independent of the other components. Elastic modulus is also a summation effect, with the upper limit $E_c(u)$ given by

$$E_c(u) = V_p E_p + V_m E_m \tag{7.9}$$

and the lower limit, $E_c(l)$ given by

$$E_c(l) = \frac{E_m E_p}{V_m E_p + V_p E_m} \tag{7.10}$$

where E_m and E_p are the elastic moduli of the matrix and particulate, respectively; and V_m and V_p are their respective volume fractions. A complementation effect occurs when each constituent contributes separate properties. For example, laminar composites are sandwich-type composites composed of several layers of materials. Sometimes the outer layer is simply a protective coating, such as a polymeric film, that imparts corrosion resistance to the composite. This outer layer serves no structural purpose, and contributes only a specific property to the overall composite—in this case, corrosion resistance. Finally, some constituent properties are not independent of each other, and an interaction effect occurs. In this case, the composite property may be higher than either of the components, and the effect may be synergistic rather than additive. For example, it has been observed that the strength of some glass-fiber-reinforced plastic composites is greater than either the matrix or the reinforcement component by itself.[5] Generally, the composite overcomes the weaknesses of constituents and its comprehensive property can be designed according to the service requirements. Therefore the composites have an extremely wide range of applications.

3) Matrix

The matrix of composite may be a metal, polymer, or ceramic. In general, metals and polymers are used as matrix materials because some ductility is desirable; for ceramic-matrix composites, the reinforcing component is added to improve fracture toughness.

The matrix serves two primary functions, to hold the fibrous phase in place and to deform and distribute the stress under load to the reinforcement phase. In most cases, the matrix material for a fiber composite has an elongation at break greater than the fiber; that is, it must deform more before breaking. It is also beneficial to have a matrix that encapsulates the reinforcement phase without excessive shrinkage during processing. A secondary function of the matrix is to protect the surface of the reinforcement. Many reinforcements tend to be brittle, and the matrix protects them from abrasion and scratching, which can degrade their mechanical properties. The matrix can also protect the reinforcement component from oxidation or corrosion. In this way, many fibers with excellent mechanical properties, such as graphite fibers, can be used in oxidizing environments at elevated temperatures due to protection by the matrix constituent.

4) Reinforcement

In reinforced composites, the reinforcement is the structural constituent and determines the internal structure of the composite. The reinforcement may take on the form of particulates, flakes, lamina, or fibers or may be generally referred to as "filler". Fibers are the most common type of reinforcement, resulting in fiber-matrix composites (FMCs).

An important characteristic of most materials, especially brittle ones, is that a small-diameter fiber is much stronger than the bulk material. The probability of the presence of a critical surface flaw that can lead to fracture diminishes with decreasing specimen volume, and this feature is used to advantage in the fiber-reinforced composites. Also, the materials used for reinforcing fibers have high tensile strengths.

On the basis of diameter and character, fibers are grouped into three different classifications: whiskers, fibers, and wires. Whiskers are very thin single crystals that have extremely large length-to-diameter ratios. As a consequence of their small size, they have a high degree of crystalline perfection and are virtually flaw free, which accounts for their exceptionally high strengths; they are among the strongest known materials. In spite of these high strengths, whiskers are not utilized extensively as a reinforcement medium because they are extremely expensive. Moreover, it is difficult and often impractical to incorporate whiskers into a matrix. Whisker materials include graphite, silicon carbide, silicon nitride, and aluminum oxide; some mechanical characteristics of these materials are given in Table 7.11.

Materials that are classified as fibers are either polycrystalline or amorphous and have small diameters; fibrous materials are generally either polymers or ceramics (i.e. the polymer aramids, glass, carbon, boron, aluminum oxide, and silicon carbide). Table 7.11 also presents some data on a few materials that are used in fiber form.

Table 7.11 Characteristics of several fiber-reinforcement materials

Material	Specific gravity	Tensile strength [GPa (10^6 psi)]	Specific strength (GPa)	Modulus of elasticity [GPa (10^6 psi)]	Specific modulus (GPa)
Whiskers					
Graphite	2.2	20 (3)	9.1	700 (100)	318
Silicon nitride	3.2	5-7 (0.75~1.0)	1.56~2.2	350~380 (50~55)	109~118
Aluminum oxide	4.0	10~20 (1~3)	2.5~5.0	700~1,500 (100~220)	175~375
Silicon carbide	3.2	20 (3)	6.25	480 (70)	150
Fibers					
Aluminum oxide	3.95	1.38 (02)	0.35	379 (55)	96
Aramid (Kevlar 49™)	1.44	3.6~4.1 (0.525~0.600)	2.5~2.85	131 (19)	91
Carbon	1.78-2.15	1.5~4.8 (0.22~0.70)	0.70~2.70	228~724 (32~100)	106~407
E-glass	2.58	3.45 (0.5)	1.34	72.5 (10.5)	28.1
Boron	2.57	3.6 (0.52)	1.40	400 (60)	156
Silicon carbide	3.0	3.9 (0.57)	1.30	400 (60)	1.33
UHMWPE (Spectra 900™)	0.97	2.6 (0.38)	2.68	117 (17)	121
Metallic Wires					
High-strength steel	7.9	2.39 (0.35)	0.30	210 (30)	26.6
Molybdenum	10.2	2.2 (0.32)	0.22	324 (47)	31.8
Tungsten	19.3	2.89 (0.42)	0.15	407 (59)	21.1

Fine wires have relatively large diameters; typical materials include steel, molybdenum, and tungsten. Wires are utilized as a radial steel reinforcement in automobile tires, in filament-wound rocket casings, and in wire-wound high-pressure hoses.

(2) Classification of composite

The composite are generally classified according to matrix, shape of reinforcement, application, etc.

1) Classification according to matrix

According to matrix, the composites can be divided into metal matrix composite, ceramic matrix composite, polymer matrix composite, etc.

2) Classification according to the shape of reinforcement

According to the shape of reinforcement, the composites can be divided into particle reinforced composite, fiber reinforced composite, laminar composite, etc.

a) Particle reinforced composite

Particle reinforced composite refers to the composite whose reinforcement is particle. $Al-Al_2O_3$ composite, Al-graphite composite, etc. are the typical particle reinforced composites.

Large-particle and dispersion-strengthened composites are the two subclassifications of particle-reinforced composites. The distinction between these is based upon reinforcement or strengthening mechanism. The term "large" is used to indicate that particle-matrix interactions cannot be treated on the atomic or molecular level; rather, continuum mechanics is used. For most of these composites, the particulate phase is harder and stiffer than the matrix. These reinforcing particles tend to restrain movement of the matrix phase in the vicinity of each particle. In essence, the matrix transfers some of the applied stress to the particles, which bear a fraction of the load. The degree of reinforcement or improvement of mechanical behavior depends on strong bonding at the particle-matrix interface.

For dispersion-strengthened composites, particles are normally much smaller, with diameters between 0.01 and 0.1 μm (10 and 100 nm). Particle-matrix interactions that lead to strengthening occur on the atomic or molecular level. The mechanism of strengthening is similar to that for precipitation hardening. Whereas the matrix bears the major portion of an applied load, the small dispersed particles hinder or impede the motion of dislocations. Thus, plastic deformation is restricted such that yield and tensile strengths, as well as hardness, improve.

b) Fiber reinforced composite

Fiber reinforced composite (shown in Figure 7.16) refers to the composite whose reinforcement is fiber. $Al-SiC_w$ composite, Al-C composite, etc. are the typical fiber reinforced composites.

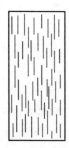  

(a) Continuous and aligned   (b) Discontinuous and aligned   (c) Discontinuous and randomly oriented

Figure 7.16   Schematic representations of fiber reinforced composites

Design goals of fiber-reinforced composites often include high strength or stiffness on a weight basis. These characteristics are expressed in terms of specific strength and specific modulus parameters, which correspond, respectively, to the ratios of tensile strength to specific gravity and modulus of elasticity to specific gravity. Fiber-reinforced composites with exceptionally high specific strengths and moduli have been produced that utilize low-density fiber and matrix materials.

Fiber-reinforced composites are subclassified by fiber length. For short fiber, the fibers are too short to produce a significant improvement in strength.

c) Laminar composite

Laminar composite refers to the composite whose reinforcement is a layer. Steel-Al composite plate, steel-Ti composite plate, etc. are the typical laminar composites.

A laminar composite is composed of two-dimensional sheets or panels that have a preferred high-strength direction such as is found in wood and continuous and aligned fiber-reinforced plastics. The layers are stacked and subsequently cemented together such that the orientation of the high-strength direction varies with each successive layer (shown in Figure 7.17). For example, adjacent wood sheets in plywood are aligned with the grain direction at right angles to each other. Laminations may also be constructed using fabric material such as cotton, paper, or woven glass fibers embedded in a plastic matrix. Thus a laminar composite has relatively high strength in a number of directions in the two-dimensional plane; however, the strength in any given direction is, of course, lower than it would be if all the fibers were oriented in that direction. One example of a relatively complex laminated structure is the modern ski.

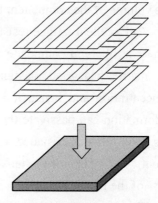

Figure 7.17   Stacking of successive oriented, fiber-reinforced layers for a laminar composite

3) Classification according to application

According to the application, composites can be divided into structural composite, functional composite, intelligent composite, etc.

(3) Thermodynamics of composite

If the composite matrix is composed of a metal, ceramic, or polymer, its phase stability behavior will be dictated by the free energy considerations. Unary, binary, ternary, and even higher-order phase diagrams can be employed as appropriate to describe the phase behavior of both the reinforcement or matrix component of the composite system.

(4) Kinetic processes in composites

Due to the fact that industrial composites are made up of combinations of metals, polymers, and ceramics, the kinetic processes involved in the formation, transformation, and degradation of composites are often the same as those of the individual components. Most of the processes we have described to this point have involved condensed phases—liquids or solids—but there are two gas-phase processes, widely utilized for composite formation, that require some individualized attention. Chemical vapor deposition (CVD) and chemical vapor infiltration (CVI) involve the reaction of gas phase species with a solid substrate to form a heterogeneous, solid-phase composite.

(5) Performance of composite

1) High specific strength and specific modulus

In general, the composite result of materials is the reduction of density. The high specific strength and specific modulus are the outstanding characteristics of composite. The higher the specific strength or specific modulus of material is, the smaller the self-weight or the volume of the component becomes.

2) Excellent fatigue resistance and fracture resistance

The fibers in composite have less defects and have high fatigue resistance. In addition, the matrix has good plasticity and toughness which can eliminate or reduce the stress concentration and hence it is not easy to generate microcracks. Furthermore, the plastic deformation can passivate the microcracks and retard their growth. Therefore the composite has good fatigue resistance. For example, the fatigue limit of carbon fiber reinforced resin is about 70% ~ 80% of the tensile strength, but the fatigue limit of metal is only about 30% ~ 50% of the tensile strength.

For composite, large amount of fine fibers distribute in the matrix. When a part of the fibers break under a large load, the load can be rapidly redistributed to the unbroken fibers through the tough matrix and hence the fracture of composite does not happen instantaneously. Therefore the composite has good fracture resistance.

3) Good high temperature resistance

High temperature resistance means that the material still maintains high strength at high temperatures. The composite has good high temperature resistance which can be briefly described as follows. The service temperature of polymer matrix composite can reach up to 350 ℃. As for metal matrix composite, the service temperature can reach up to 1,100 ℃. SiC fiber and Al_2O_3 fiber ceramic composites can maintain high strength in the range of 1,200 ~ 1,400 ℃. The carbon fiber composite can be used for a long period of time at 2,400 ~ 2,800 ℃ in a non-oxidizing atmosphere.

4) High friction and wear property

The composite has good friction and wear property. Its friction coefficient is lower than that of polymer materials and the fibers improve the wear resistance greatly.

5) Good vibration damping performance

The composite has strong damping capacity, high specific elastic modulus and self-vibration frequency. Furthermore, the interface between fibers and matrix can absorb the vibration energy and attenuate the vibration quickly. Therefore, it is not easy to resonate for the part.

6) Other special features

Metal matrix composite has high toughness and thermal shock resistance. Glass fiber reinforced plastic has excellent electrical insulation. Electricity and magnetism have no effect on it. Furthermore, it cannot reflect the radio wave. In addition, composite also has good radiation resistance, high creep performance and perfect optical, electrical and magnetic properties.

(6) Strengthening mechanism and rules of composite

1) Fiber reinforced composite

a) Strengthening mechanism

In the fiber reinforced composite, the fibers are used as reinforcements. The materials used to process fibers commonly have strong bond and high hardness such as ceramic, glass, etc. These materials contain a lot of inner microcracks and have great brittleness when they are in block shape. The use of fine fibers can reduce the length and quantity of crack i.e. reduce the brittleness. Accordingly the role of reinforcements can be exerted as follows:

- The reinforcements in the polymer matrix composite can effectively impede the movement of matrix molecular chain.
- The reinforcements in the metal matrix composite can effectively inhibit the dislocation motion and hence strengthen the matrix.

b) Rules of composite

The rules of composite are as follows:

- The reinforcement is the main carrier of load and should have high strength and elastic modulus higher than those of the matrix.
- The matrix with certain plasticity and toughness acts as a binder and should have wettability to the reinforcement.
- The difference of thermal expansion coefficient between matrix and reinforcement should be small enough.
- The reinforcement must have a reasonable content, size and distribution.
- Harmful chemical reactions between matrix and reinforcement must be avoided.

2) Particle reinforced composite

a) Strengthening mechanism

In the particle reinforced composite, the matrix with certain plasticity and toughness bears load. The particles strengthen the matrix by impeding the movement of matrix molecular chains and dislocations. The enhanced effect is closely related to the volume content, distribution and size of particle.

b) Rules of composite

The rules of composite are as follows:
- In order to hinder the movement of molecular chains or dislocations effectively, the particles should be uniformly dispersed in the matrix.
- The particle size should be appropriate. The too large particles are liable to fracture and can cause the stress concentration resulting in the reduction of material strength. However, the too small particles cannot impede the movement of dislocations and the strengthening effect is not ideal. Generally, the particle diameter is from several micrometers to several tens of micrometers.
- The volume fraction of particles should be above 20%, otherwise the best strengthening effect will not be achieved.
- There should be a certain bonding strength between the particles and the matrix.

(7) Processing of metal matrix composite

1) Development of metal matrix composite

Metal matrix composite really originated in the late 1950s or early 1960s. The research and development of metal matrix composites starts with the successful production of W-reinforced copper matrix composite by the National Aeronautics and Space Administration (NASA).

Subsequently, the research on fiber reinforced metal matrix composites developed rapidly in the 1960s. The researches mainly focused on the aluminum and copper composites reinforced by tungsten and boron fibers. In these composites, the primary function of matrix is to transmit and distribute the load to the fibers. The volume fraction of reinforcements is generally very high (about 40% ~ 80%). Therefore the axial performance is very good and the microstructure and strength of the matrix seem to be subordinate.

The researches on the continuous fiber reinforced composites was slow in the 1970s mainly due to the high price of raw materials and the limitation of manufacturing manners. The demand for the high temperature materials used in turbine engine components has triggered the widespread interest in metal matrix composites, especially in titanium matrix composites.

Because the metal matrix composites have extremely high specific strength, specific stiffness and high-temperature strength, this kind of materials is firstly applied in the aerospace field and will certainly occupy an important position in this field in the future. In addition, it has also been applied in the fields of automobile, sports, etc. In particular, the whisker reinforced composites and particle reinforced composites have been well applied in the civil field.

The key points of researches on the metal matrix composite are as follows:
- Strengthening effect of matrix and reinforcements and the design and performance of composites;
- Optimization and design of interface between matrix and reinforcements;
- Preparation process to improve the performance and reduce the cost of composites;
- New enhancer;
- Application of composites.

2) Interfacial bonding form of metal matrix composite

The interfacial bonding form of metal matrix composite includes interfacial mechanical combination, interfacial infiltration and dissolution, interfacial chemical reaction combination, etc.

a) Interfacial mechanical combination

Interfacial mechanical combination primarily relies on the mechanical "anchor" force on the rough surface of reinforcement.

b) Interfacial infiltration and dissolution

Interfacial infiltration and dissolution mainly relies on the interfacial combination resulted from the infiltration and dissolution of reinforcement on the interface. If the mutual dissolution is serious, the precipitation phenomenon may occur after dissolution, damage the enhancement seriously and reduce the interfacial bonding properties of the composite.

c) Interfacial chemical reaction combination

Interfacial chemical reaction combination mainly relies on the interfacial chemical reactions. A chemical potential gradient between the matrix and the reinforcement exists at the interface for most metal matrix composites. Inter-diffusion and chemical reaction can occur as long as there is a favorable kinetic condition.[6] From the interface optimization, the interfacial reaction between reinforcement and matrix can take place properly after wetting and thus the interfacial bonding strength of composite can be further improved.

For all the interfacial bonding forms, improving the wettability of reinforcement to the matrix, controlling the speed of interfacial reaction and the quantity of reactants, eliminating the interfacial layer which can seriously worsen the performance of composite and optimizing the interface design of composite are the important parts of researches on the metal matrix composites interface.

3) Interface optimization and interface design

Interface optimization and interface design generally can be carried out by the following ways.

a) Modifying the surface of reinforcement

The commonly used methods for the surface modification treatment of reinforcements include PVD, CVD, electrochemistry method, sol-gel method, etc.

b) Improving the mechanical properties of reinforcement

The commonly used method for the improvement of reinforcement mechanical property is diffusion annealing.

c) Improving the wettability of reinforcement to the matrix

By adding 3% magnesium as an active element to aluminum, the surface energy of the liquid aluminum can be lowered and thereby the wettability of reinforcement to the matrix can be improved.

d) Avoiding the diffusion, penetration and reaction between reinforcement and matrix

By adding alloying elements Al, Mo, V, Zr, etc. to pure titanium can significantly reduce the reaction rate between titanium alloy and boron fiber.

e) Reducing the difference of elastic modulus and thermal expansion coefficient between reinforcement and matrix.

An appropriate transition layer on the reinforcement can reduce the difference of elastic modulus and thermal expansion coefficient between reinforcement and matrix.

4) Classification of preparation process of metal matrix composite

The common preparation process of metal matrix composite includes solid phase method, liquid phase method, in-situ synthesis method, etc.

Various ceramics can be combined with metals to form composites with different special performance by new technology. It can be believed that ceramics and composites will play a more important role in wide fields such as rail traffic, aviation, navigation, aerospace, etc.

Chapter 7　Materials Science and Technology
第7章　材料科学与工程

Words and Expressions

flaw [flɔː] an imperfection in a device or machine	n. 缺陷
optic fiber [ˈɒptɪk ˈfaɪbə]	光纤
nano ceramic [ˈnænəʊ səˈræmɪk]	纳米陶瓷
matrix [ˈmeɪtrɪks] a rectangular array of elements (or entries) set out by rows and columns	n. 基体
reinforcement [ˌriːɪnˈfɔːsmənt] the act of making sth. stronger	增强相
interphase [ˈɪntəfeɪz] interphase is a 1989 3D first-person and puzzle video game developed by The Assembly Line and published by Image Works for multiple platforms.	中间相
combination effect [ˌkɒmbɪˈneɪʃən ɪˈfekt]	互补效应
in essence [ɪn ˈesns]	本质上
laminar composite [ˈlæmɪnə ˈkɒmpəzɪt]	层状复合材料
free energy [friː ˈenədʒi]	自由能
unary [ˈjuːnəri] consisting of or involving a single element or component	adj. 一元的
binary [ˈbaɪnəri] a system of two stars that revolve around each other under their mutual gravitation	n. 二元的
ternary [ˈtɜːnəri] the cardinal number that is the sum of one and one and one	n. 三元的
subordinate [səˈbɔːdɪnət] an assistant subject to the authority or control of another	n. 次要的
inter-diffusion [ɪntəː dɪfˈjuːʒən]	互扩散

Notes

[1] 本句中 it 为形式主语，全句可译为：然而，这种孔隙的消除过程进行得不完全，通常残留一定的孔隙。

[2] as a consequence 可译为"因此"。

[3] 本句为 not only … but also 句型，全句可译为：耐火陶瓷不仅能够承受高温、隔热，而且在许多情况下，还必须抵抗熔融金属、盐、炉渣和玻璃的高温侵蚀。

[4] 本句含较多并列成分，全句可译为：无论在需要多孔部件时、部件形状复杂时、任何其他方式不能混合时或者混合物熔点非常高时，都能使用粉末冶金。

[5] 本句为 either…or 句型，全句可译为：例如，已经观察到一些玻璃纤维增强塑料复合材料的强度既大于基体又大于增强相。

[6] 本句为让步状语从句，全句可译为：只要存在有利的动力学条件，就可以发生互扩散和化学反应。

Questions for Discussion

1. What is the static fatigue?
2. Why does the porosity deteriorate the flexural strength?
3. What is sintering?
4. What are the influence factors of sintering?
5. Which unique performance does the new ceramic possess?
6. How does the fiber can be divided? What are they?
7. How does the process of chemical vapor deposition can be divided? What are they?
8. What are the performance of composite?
9. What is the strengthening mechanism of fiber reinforced composite?
10. Which rules are there in particle reinforced composite?
11. What are the key points of researches on the metal matrix composite?
12. How do the interface optimization and interface design conduct?
13. How many preparation processes are there in metal matrix composite manufacturing?

7.3　Brake Disc of Multiple Units

"Multiple units" is a type of modern train consisting of at least two powered carriages and several non-powered carriages. Brake disc is a key part to ensure the safe and reliable braking performance of multiple units.

7.3.1　Brake Disc Overview

(1) Definition of brake disc

The brake disc is a disc-shaped moving part which fixes on the wheel and can stop the rotation of wheel by the friction between its end plane and the friction element (shown in Figure 7.18).

Figure 7.18   Brake disc

(2) Advantages of brake disc

With the increase of vehicle speed and total weight, the disc brake with good thermal stability gradually replaces the drum brake. The main advantages of brake disc are as follows:

- Brake disc has better heat dissipation than brake drum and it will not cause brake recession and brake failure in the case of continuous braking.
- The change in size of the brake disc after heating does not increase the stroke of the brake pedal.
- The brake disc system can react quickly and make high frequency brake action. It is easy to install the ABS and other control systems.
- Brake disc does not have the automatic brake action of brake drum, so the brake force of left and right wheels is relatively average.
- Brake disc has better drainage ability and can reduce the bad braking resulting from the water or sediment.
- Brake disc is simple in construction and easy to repair.

(3) Common type of brake disc

1) Classification according to assembly mode

According to the assembly mode, the common brake disc can be divided into wheel mounted brake disc and axle mounted brake disc. Figure 7.19 shows the forged steel wheel mounted brake disc and axle mounted brake disc used in the CRH2[1] multiple units.

(a) Forged steel wheel mounted brake disc   (b) Forged steel axle mounted brake disc

Figure 7.19 Brake disc mode

2) Classification according to composition of brake disc

According to the composition of brake disc, brake disc can be divided into iron carbon alloy brake disc and composite brake disc. As the brake disc is a key part of the train braking system, the material of brake disc will play a crucial role in determining the performance of brake disc[2].

(4) Brake mode of brake disc

The braking mainly includes three cases: ① The breaking in slowing down or even stopping of the moving vehicle according to the requirements of driver. ② The breaking in stabilizing the parked vehicles. ③ The breaking in keeping the speed of downhill vehicle steady.

In all kinds of vehicle braking systems, the brake is the part used to produce the force against the motion or motion trend of vehicle. At present, the brake used in all kinds of vehicles is frictional brake, that is, the brake torque preventing the vehicle from moving is caused by the friction between the fixed element and the rotating working surface.

At present, no matter in Shinkansen of Japan, TVG of France, ICE of Germany or in CHR of China[3], the braking mode is inseparable from the combination of dynamic braking and disc braking[4]. The relevant situations are shown in Table 7.12.

Table 7.12 Braking mode of high speed train

Train model	Maximum speed (km/h)	Braking mode
300	300	Dynamic+ disc
ICE	300	Dynamic+disc+electromagnetic rail
TGV-A	300	Dynamic+ disc
TGV-PSE	270	Dynamic+ disc+tread
CHR	300	Dynamic+disc+electromagnetic rail
CHR	200	Dynamic+ disc

(5) Performance requirements for brake disc material

Brake disc is the key part of brake and plays an important role in the safety of high speed train. It serves in the harsh friction braking of high speed trains. Therefore, the brake disc materials must meet the following requirements:

• Sufficient strength to withstand the centrifugal force of high speed rotation and brake

disc pressure;
- High and stable friction coefficient to obtain good braking effect;
- High wear resistance to reduce wear caused by intense friction between disc and brake;
- Good thermal cracking resistance to hinder the crack of brake disc under shock hot and cold conditions;
- Low density to reduce the mass of vehicle;
- Good process performance;
- Low cost.

7.3.2 Iron Carbon Alloy Brake Disc

The iron carbon alloy brake disc can be divided into iron brake disc and steel brake disc. The common ones include cast iron brake disc, cast steel brake disc, forged steel brake disc, cast iron-cast steel composite brake disc and so on.

(1) Cast iron brake disc

The carbon content of cast iron brake disc is about 3.5 wt% and its density is about 7.2 g/cm^3. According to the graphite shape and alloy element content, the common cast iron brake disc includes flake graphite cast iron brake disc, Ni-Cr-Mo low-alloy cast iron brake disc and vermicular graphite cast iron brake disc.

Iron carbon alloy brake disc

1) Flake graphite cast iron brake disc

The composition of flake graphite cast iron brake disc is shown in Table 7.13. It can be seen that the carbon content is higher about 3.5 wt%. In the flake graphite cast iron brake disc, the carbon exists in the form of sheet graphite, resulting in the following characteristics of brake disc:
- Stable friction coefficient about 0.25~0.35;
- Good wear resistance;
- Little deformation;
- Good casting performance;
- Low cost.

Table 7.13 Composition of flake graphite cast iron brake disc

Material of brake disc	Composition (wt %)								
	C	Si	Mn	P	S	Ni	Cr	Mo	Fe
Flake graphite cast iron	3.2~3.9	1.5~2.1	0.6~1	<0.16	<0.12	-	-	-	Balance
Ni-Cr-Mo low-alloy cast iron	3.3~3.7	1.1~1.6	0.6~1	<0.16	<0.12	1~2	0.3~0.6	0.3~0.5	Balance

However, this brake disc can be badly worn at high speed and needs frequent replacement. Thus it is rarely used in high speed trains nowadays.

2) Ni-Cr-Mo low-alloy cast iron brake disc

In the smelting process of iron alloy, some alloying elements (W, Cr, V, Mo, Ni, Mn, Ti) can be added to improve its special properties. After adding Ni, Cr and Mo, the flake graphite cast iron brake disc turns into the Ni-Cr-Mo low-alloy cast iron brake disc whose composition is shown in Table 7.13. Due to the addition of Ni, Cr and Mn, the thermal shock resistance and the high-temperature wear resistance of Ni-Cr-Mo low-alloy cast iron brake disc are better than those of flake graphite cast iron brake disc. However, its strength and thermal fatigue properties are weak and thus the thermal cracks are easy to occur, resulting in its failure at high speed.

3) Vermicular graphite cast iron brake disc

The vermicular graphite cast iron is obtained after the vermicularizing treatment to cast iron. The vermicularizing treatment is such a treatment that the flake graphite turns into the vermicular graphite after adding the vermicular agent into cast iron and thus the good thermal conductivity can be maintained and the strength and thermal shock resistance can be improved (shown in Figure 7.20 and Table 7.14).

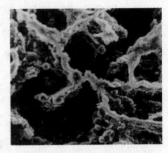

Figure 7.20 Microstructure of vermicular graphite cast iron

The vermicular agent mainly includes Mg and rare earth. So far, all researches agree that rare earth is the dominant element in making vermicular cast iron. China is rich in rare earth resources. This provides the extremely favorable conditions and material basis for the development of vermicular cast iron.

Table 7.14 Properties of vermicular graphite cast iron

Brand	Yield strength (MPa) ≥	Tensile strength (MPa) ≥	Elongation (%) ≥	Hardness (HBS)	Vermicularity (%) ≥
RuT420	335	420	0.75	200~280	50
RuT380	300	380	0.75	193~274	
RuT340	270	340	1.0	170~249	
RuT300	240	300	1.5	140~217	
RuT260	195	260	3	121~197	

The vermicular graphite cast iron brake discs have been used on high speed railways in Japan, Britain and the United States. This brake disc is also developed by the vehicle technology institute of Qishuyan in China and the loading test has been carried out.

(2) Cast steel brake disc

Carbon content of cast steel brake disc with a density of 7.8 g/cm^3 is generally less than 0.3 wt%. Besides, it also contains Cr, Mo and V (shown in Table 7.15). It has excellent performance in strength, impact toughness and thermal shock resistance (shown in Table 7.16). ICE of Germany has used it before. However, this kind of brake disc has several defects in thermal conductivity, thermal capacity and thermal expansion coefficient. Furthermore, the groove, wrinkle, mesh crack and local melting in the friction surface greatly aggravate the wear of brake disc and pad. Therefore it is being replaced by the forged steel disc.

Table 7.15 Composition of cast steel brake disc

Brand	C	Cr	Mo	V	Fe
28CDV5.08	0.24~0.31	1.2~1.6	0.6~0.9	0.2~0.4	Balance
15 CDV.10	0.15	1~1.5	0.85~1.15	0.15~0.3	Balance

Table 7.16 Properties of cast steel brake disc

Brand	Ultimate strength (MPa)	Tensile strength (MPa)	Elongation (%)	Impact energy (J)	Hardness (HB)
28CDV5.08	1,050~1,250	>970	>10	>23	331~388
15 CDV.10	1,030~1,200	>900	>10	>20	331~388

(3) Cast iron-cast steel composite brake disc

Cast iron-cast steel composite brake disc has attracted people's attention in recent years. For this brake disc, the friction disc spoke is made of cast iron and the disc hub is made of cast steel (shown in Figure 7.21). In this way, the stable and good friction performance of cast iron and the thermal shock resistance, high strength and good elongation of cast steel can be utilized. Furthermore, compared with the cast iron brake disc, the service life of cast iron-cast steel composite brake disc is increased two times and the cost is lowered remarkably. Therefore, the cast iron-cast steel composite brake disc is loved by various countries and is used on Japan Shinkansen at one time.

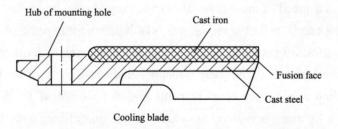

Figure 7.21 Structure of cast iron-cast steel composite brake disc

(4) Forged steel brake disc

The composition of forged steel brake disc is shown in Table 7.17. Forged steel brake disc is subjected to a plastic deformation process and cast steel brake disc is subjected to a liquid forming process. Compared with cast steel brake disc, forged steel brake disc possesses better performance in strength, hardness, impact resistance, etc. (shown in Table 7.18) Moreover, forged steel brake disc has high thermal shock resistance, good wear resistance and thermal fatigue resistance, resulting in a longer service life. Therefore, it has been used on Shinkansen, ICF and TGV-A high speed trains.

Table 7.17　Composition of forged steel brake disc (wt %)

C	Si	Mn	P	S	Ni	Cr	V	Mo	Fe
0.2~0.6	0.2~0.8	0.3~2	<0.03	<0.03	0.3~2	0.4~2.5	<0.35	0.2~1	Balance

Table 7.18　Properties of forged steel brake disc

Yield strength (MPa)	Tensile strength (MPa)	Elongation (%)	Reduction of area (%)	Impact energy (J)	Hardness (HBS)
>1,165~1,240	1,255~1,330	>13~13.5	53.5~54	>42~95	363~378

7.3.3　Composite Brake Disc

Composite brake disc

(1) Introduction of composite brake disc

Under the general trend of high speed and heavy load, it is indispensable to reduce the self-weight of the train and improve the speed of the train. In this case, cast iron or cast steel brake discs cannot meet the braking requirements of the train because their high density (>7g/cm^3) limits the increasing of train speed greatly. Take Japan's Shinkansen as an example, four steel brake discs are installed on each axle and thus one bogie has eight brake disks whose total weight is about 0.5 tons which accounts for about 20 percent of the train's unsprung weight, while the ideal braking disc weight for high speed trains should be about 10 percent of the unsprung weight[5]. On the other hand, the traditional cast iron brake disc cannot meet the braking requirements of the high speed train due to its large tendency of thermal cracking at a speed of more than 200 km/h. Forged steel brake disc can still meet the requirements, but it cannot meet the requirements of lightweight because of its large density.

The new composite brake disc is a beneficial supplement to ferro-carbon alloy brake disc. For composite brake disc, the wear resistance and service life of the brake disc can be improved by adding appropriate reinforcement and the weight of the brake disc can be reduced by using a lightweight matrix. For example, the glass fibers with an elastic modulus of 70 GPa and tensile strength of 3,500 MPa and the carbon fibers with a modulus of 60 GPa and tensile strength of 3,900 MPa have been widely used to enhance the wear resistance of brake disc and the lightweight Al alloy has been used for the matrix of brake disc. According to the mixture law, researchers can process the light composite brake disc with good properties.

(2) Characteristics of composite brake disc

The characteristics of composite brake disc are as follows:

- The brake disc made of composite material is light in weight, high in strength and strong in rigidity.
- The production efficiency of composite brake disc is high and the manufacturing cost is low owing to its near net shape forming without additional mechanical processing[6].
- The composite brake disc can improve the safety and comfortability, decrease the noise and vibration and reduce maintenance.
- High fatigue strength and low notch sensitivity[7] of the composite brake disc ensure the high safety of train.
- Low cost.
- With the development of raw material and technological progress, its cost is gradually reduced.
- The modular design technology and integral molding technology greatly reduce the complexity of structure, shorten the actual production and final assembly work and thus reduce the overall cost.
- Due to the good properties in fatigue and corrosion, the cost of maintenance is reduced.
- As a result of the reduction in weight, the operational capacity is increased, the energy consumption is reduced and thus the long run total cost is reduced.

(3) Type of composite brake disc

The new type of composite brake disc includes carbon / carbon fiber composite brake disc, ceramic reinforced composite brake disc, surface coated brake disc and aluminum matrix composite brake disc.

1) Carbon / carbon fiber composite brake disc

Carbon / carbon fiber composites are carbon fiber reinforced carbon matrix composites. A carbon fiber reinforced graphite composite brake disc has high specific heat value, low thermal expansion coefficient, little elastic modulus and good thermal conductivity and thermal shock resistance. In particular, it can work normally under the high temperature of 1,000 ℃ and the highest working temperature can reach up to 2,000 ℃ which is impossible for other brake discs. In addition, the density of this composite brake disc is 1.75 g/cm^3, one fifth of that of iron, which creates the condition for reducing weight. However, its friction coefficient varies greatly with velocity, pressure and humidity especially in the case of rain or snow, resulting in an increase of wear loss. Furthermore, the high temperature of the brake disc puts forward a harsh requirement on the adjacent accessories. In addition, the factors such as its complex structure and high price of expensive carbon fiber are impeding the extensive application of carbon / carbon fiber composite brake disc. Fortunately, its excellent performance has attracted more and more attention and Japan, Germany and France are developing this brake disc.

2) Ceramic reinforced composite brake disc

Ceramics with high strength, good high temperature performance and wear resistance might be used for anti-friction devices. If the brake made of ceramic material is successfully applied on high speed trains, the total weight of the brake disc on each bogie can be reduced from 1,560 kg to 750 kg. But their brittleness is a major obstacle to prepare brake discs for high speed trains.

At present, ceramic reinforced composite brake disc has been developed. By means of plasma thermal spraying technology, a carbide ceramic coating rich in Ni and Cr is manufactured on the surface of steel brake disc. The coating has a hardness of 800 HV and a thickness of 250 μm. The ceramic reinforced composite brake disc can cause a higher surface temperature and a violent abrasion at the contact surface, the ordinary powder metallurgical brake pad cannot work competently in these conditions. Therefore, the matching brake pad is also reinforced by ceramic and has a hardness of 420 HV.

The results show that the friction coefficient between the ceramic reinforced steel brake disc and the ceramic reinforced brake pad is quite stable. In particular, under the condition of water, the friction coefficient is almost unchanged. This feature is far better than that between the non-reinforced steel brake disc and copper powder metallurgical brake pad. The reason why the friction coefficient between ceramics is not affected by moisture is that the high temperature of friction surface makes it difficult to form a water film[8].

3) Surface coated brake disc

It is a new way for processing a high quality brake disc to generate a surface layer which can improve the friction and wear performance. Because the brake disc has a high demand on both thermal and mechanical properties, it can improve the performance of the brake disc through surface strengthening technology. Generally, two aspects are considered. One is to improve the mechanical property of brake disc surface only. The other is to enhance the thermal property on the basis of the improved mechanical property. These technological thoughts can be realized mainly by plasma thermal spraying technology and thermal spraying technology.

4) Aluminum matrix composite brake disc

The aluminum alloy cannot meet the requirements of friction and wear when the train is braking. In order to solve this problem, the aluminum matrix composite brake disc is made by adding ceramic particles in aluminum alloy matrix and thus the friction and wear properties are improved.

The application of aluminum matrix composite brake disc can reduce the weight by more than 40 %. Aluminum matrix composite brake disc possesses good heat dissipation capacity and has a heat dissipation capacity of 4,200 W (shown in Figure 7.22). The simulation results show that Al-SiC composite brake disc can absorb energy up to 18 MJ under a 120 s maintaining braking.

Figure 7.22   Aluminum matrix composite brake disc

For Al-ceramic composites, the ceramic particles mainly include Al_2O_3 or SiC particles. Their manufacturing manner is semisolid casting technology, that is, the ceramic particles are firstly stirred into Al slurry by mechanical mixing method and then get the aluminum matrix composite brake disc by casting.

The comparison test with steel disc shows that the friction coefficient of aluminum-SiC composite brake disc and synthetic brake pad is stable around 0.4, higher than that of steel brake disc and synthetic brake pad. When the pressure is higher than 0.3 MPa, the friction coefficient is still higher than 0.2. Under the condition of water spraying, the friction coefficient is still higher than that of steel disc and synthetic brake pad.

So far, the high-speed trains mainly use iron carbon alloy brake disc. The carbon / carbon fiber composite brake disc possesses low density and high braking energy, but its stability of friction coefficient and the high manufacturing and operating costs are still a problem. The ceramic reinforced composite brake disc has excellent high-temperature friction and wear properties, but the brittleness problem of large size parts is still not solved. The surface coated brake disc can greatly improve the friction and wear performance of the brake disc, but the problem of bonding reliability between layers needs to be solved urgently. The aluminum matrix composite brake disc can reduce more than 50 percent of weight, but its relatively high friction coefficient limits its wider use.

Words and Expressions

brake disc [breɪk dɪsk]	制动盘
end plane [end pleɪn]	端平面
brake pad [breɪk pæd]	制动闸片
heat dissipation [hiːt ˌdɪsɪˈpeɪʃən]	散热
pedal [ˈpedəl] a flat bar on a machine	n. 踏板

thermal cracking resistance ['θɜːməl 'krækɪŋ rɪ'zɪstəns]	抗热裂性
cast iron brake disc [kɑːst 'aɪən breɪk dɪsk]	铸铁制动盘
cast steel brake disc [kɑːst stiːl breɪk 'dɪsk]	铸钢制动盘
forged steel brake disc ['fɔːdʒd stiːl 'breɪk 'dɪsk]	锻钢制动盘
cast iron-cast steel composite brake disc [kɑːst 'aɪən kɑːst stiːl 'kɒmpəzɪt dɪsk]	铸铁-铸钢复合制动盘
flake graphite cast iron brake disc [fleɪk 'græfaɪt kɑːst 'aɪən breɪk dɪsk]	片状石墨铸铁制动盘
friction coefficient ['frɪkʃən ˌkəʊə'fɪʃənt]	摩擦系数
vermicular graphite cast iron brake disc [vɜː'mɪkjʊlə 'græfaɪt kɑːst aɪən breɪk dɪsk]	蠕墨铸铁制动盘
vermicularizing treatment [və'mɪkjʊləraɪzɪŋ 'triːtmənt]	蠕化处理
thermal conductivity ['θɜːməl ˌkɒndʌk'tɪvɪti]	导热性能
thermal fatigue resistance ['θɜːməl fə'tiːg rɪ'zɪstəns]	耐热疲劳性
thermal expansion coefficient ['θɜːməl ɪk'spænʃən ˌkəʊə'fɪʃənt]	热膨胀系数
thermal shock resistance ['θɜːməl ʃɒk rɪ'zɪstəns]	抗热震性能
ceramic reinforced composite brake disc [sə'ræmɪk ˌriːɪn'fɔːst 'kɒmpəzɪt breɪk dɪsk]	陶瓷强化复合制动盘
surface coated brake disc ['sɜːfɪs 'kəʊtɪd breɪk dɪsk]	表面涂层制动盘
aluminum matrix composite brake disc [ə'luːmɪnəm 'meɪtrɪks 'kɒmpəzɪt breɪk dɪsk]	铝基复合制动盘
carbon / carbon fiber composite brake disc ['kɑːbən 'faɪbə 'kɒmpəzɪt breɪk dɪsk]	碳/碳纤维复合制动盘
carbide ceramic coating rich in ... ['kɑːbaɪd sə'ræmək 'kəʊtɪŋ rɪtʃ ɪn]	富含……的碳化物陶瓷涂层

Chapter 7　Materials Science and Technology
第 7 章　材料科学与工程

powder metallurgical brake pad ['paʊdə 'metə'lɜːdʒɪkl breɪk pæd]	粉末冶金陶瓷闸片
plasma thermal spraying technology ['plæzmə 'θɜːməl 'spreɪɪŋ tek'nɒlədʒi]	等离子热喷涂技术

Notes

[1] CRH：China Railway High-speed，指中国铁路高速列车。CRH2 是构造速度在 200 km/h 至 350 km/h 之间，该系列车型是国内大功率动车组的主力军车辆在高速、启停、安全、检测、耐寒、抗沙和卧铺等方面均运用广泛，城际列车、长途列车、高速列车和高速综合检测列车等都能看到其身影，后续很多国产高速列车亦以它为基础技术平台研制。

[2] 本句为原因状语从句，全句可译为：由于制动盘是列车制动系统的关键部件，所以制动盘的材料在确定制动盘性能中起着至关重要的作用。

[3] 本句为让步状语从句，全句可译为：目前，无论是日本的新干线、法国的 TVG、德国的 ICE，还是中国的 CHR，制动模式都与动力制动和盘式制动的结合密不可分。

[4] Shinkansen 是指新干线，贯通日本的高速铁路系统，其首条线路于 1964 年开通运行，不仅是当今世界上先进的高速铁路系统，还是世界上最早进行旅客运输的高速铁路系统。可以说，诞生于 20 世纪下半叶的日本新干线，是世界高速铁路的先驱，和法国 TGV、德国 ICE 一起，并列为世界高铁三巨头；TGV 是法国高铁的法语缩写，也是法铁的注册商标；ICE 是 Intercity-Express 的缩写，是德国铁路最快及最高等级的列车类型。

[5] 本句由多个从句构成，全句可译为：以日本新干线为例，每根车轴上安装了四个钢质制动盘，一台转向架有 8 个制动盘，重量约为 0.5 t，约占列车簧下重量的 20%，而高速列车理想的制动盘重量应为簧下重量的 10%左右。

[6] near net shape forming without additional mechanical processing 是指近终成形，金属材料的近终成形是集金属合成、精炼、凝固、成形于一道工序的一次成形技术，它同时实现了减少工序、缩短生产周期、提高金属利用率及提高金属性能。目前已使用的金属近终成形技术有近终形连铸、粉末冶金、喷射沉积成形、电磁铸造等。

[7] notch sensitivity 是指缺口敏感度，材质表面的一些缺陷，如划痕、裂纹、脱碳等，都使其强度、塑性等性能发生很大变化，通常用光滑试样的抗拉强度和有缺口试样的抗拉强度的比值作为缺口敏感性指标

[8] 本句为同位语从句，全句可译为：陶瓷间的摩擦系数不受水气影响的原因在于摩擦表面的高温使水膜难于形成。

Questions for Discussion

1. What is the brake disc?

2. What advantages does the disc brake possess?
3. How does the brake disc can be divided? What are they?
4. What requirements do the brake disc materials must meet?
5. How does the common iron carbon alloy brake disc can be divided?
6. What characteristics does the flake graphite cast iron disc brake possess?
7. What are the functions of alloying elements in the smelting process of iron alloy?
8. What is the vermicularizing treatment?
9. What characteristics does the composite brake disc possess?

7.4 Liquid Die Forging and Semi-Solid Forming

7.4.1 Liquid Die Forging

(1) Basic concept of liquid die forging

Liquid die forging is such a processing method that can operate as follows: After pouring the metal melt into the high-strength pressure chamber or die cavity, a mechanical static pressure is continuously exerted to realize the filling, solidifying and rheological feeding of metal liquid under pressure. After unloading, a workpiece with a dense inner structure, a fine crystal grain, a smooth appearance, an accurate size and excellent mechanical properties can be ejected (shown in Figure 7.23).

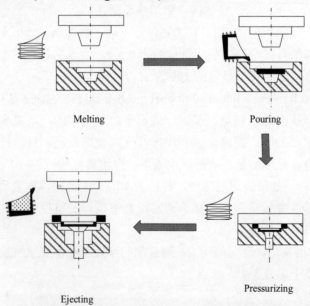

Figure 7.23   Sketch map of liquid die forging

Liquid die forging, which can maximizes the technical advantages of traditional casting and forging, is also called liquid forging or squeeze casting, and is widely used in the liquid processing of metallic materials. Its process chart is shown in Figure 7.24.

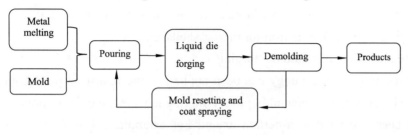

Figure 7.24　Process chart of liquid die forging

(2) Development of liquid die forging

1) Origination of liquid die forging

In 1937, the researchers combined the casting with forging and a novel liquid die forging. The scholars who are engaged in forging call it liquid die forging and those who are engaged in casting called it squeeze casting. It is first applied in the manufacturing of metal components in the military and high-tech fields. Since the liquid die forging has a comprehensive advantage of casting and forging, that is, simple, low energy consumption and high quality of workpiece, it is accepted and utilized by more and more scholars and manufacturers. Notwithstanding its decades of history, the commercialization of liquid die forging began to take place in North America, Europe, and Japan only after 1960 for the production of aluminium automotive components[1]. Thereafter, copper alloys, stainless steel, Ni-based superalloys as well as cast iron have also been used. The most remarkable application of the process was in the production of alloy wheels performed by Toyota in 1979. Recently, the liquid die forging technique has also been applied for manufacturing metal matrix composites and magnesium alloys.

2) Application of liquid die forging

In China, liquid die forging has been researched since 1958. It was put to use in the production of aluminum alloy parts in the mid 1960s. In the 1970s, liquid die forging developed rapidly. It was used to make the large aluminum alloy pistons, nickel brass high-pressure valve bodies, copper alloy worm gears, etc. After the 1980s, the process has got a great development in ferrous metal filed and the liquid die forging technology for the production of steel flat flange is developed.

Especially, after developing a special liquid die forging machine in Japan, liquid die forging has been applied in the fields of machinery and transportation in America, Japan, Britain, Germany etc. after the 1980s.

To further improve the casting quality, liquid die forging has been combined with vacuum die-casting. Aluminium alloys can be used for brake rotors, automotive and truck wheels, steering column housings, structural automotive components, pots, etc. Other alloys can also be used, such as brass and bronze for bushings, steel for missile components and differential pinion gears, ductile iron for mortar shells, etc.

3) Current problem of liquid die forging

Liquid die forging technology has been applied in the industrial production. However, due to the lack of special or general liquid die forging machine, the development of liquid die forging has been restricted to a certain extent. For example, nearly 80% of the liquid die forging machines in China are the modified hydraulic presses and the production efficiency and product quality are lower. Therefore, the most critical issue in the development of liquid die forging is to develop the special liquid die forging machines as soon as possible.

4) Development trend of liquid die forging

Different process parameters influence the quality of liquid die forging parts; in particular, pressure and die temperature appear to be the most important. In order to optimize such parameters, process simulation provides a powerful tool to rapidly find the appropriate casting conditions and also to assist in runner and gating system design. However, not many papers discuss this application due to the limits of foundry simulation software in managing element compression generated by hydrostatic pressure during filling and solidification.

In the future, the development of liquid die forging will focus on the diversification of liquid die forging to exploit the advantage of liquid forging and overcome the defects of casting and forging, e.g. liquid die forging of ceramic reinforced metal matrix composites, liquid die forging of high silicon aluminum alloys, liquid die forging of cast iron, etc. to realize the casting of high silicon aluminum alloys and the forging of cast iron.

(3) Advantages and disadvantages of liquid die forging

1) Advantages

a) High density and mechanical properties of products

Since the liquid die forging uses the external mechanical pressure to achieve the rheological feeding, it is easier to obtain a dense cast than the ordinary gravity casting which achieves the feeding by the gravity of riser. Furthermore, liquid die forging also uses lower ram velocities than die-casting (about 0.5 m/s) and the liquid metal is introduced into the die with very low turbulence; this strongly reduces the risk of air entrapment and porosity formation. The pressure applied before, during, and after alloy solidification guarantees contact between metal and die walls, increasing heat flow rate and, thus, cooling rate, giving rise to a fine microstructure[2]. Thus the work pieces of liquid die forging usually have no shrinkage holes and shrinkage defects existing in the conventional casts. In addition, since nucleation of gas porosity depends on the pressure, the porosity formation due to dissolved gases is restricted. The result is a denser, fine-grained, weldable, and heat-treatable cast, characterized by excellent mechanical properties and superb surface without fin lines. The mechanical properties of workpieces are shown in Table 7.19.

Table 7.19 Typical mechanical properties of liquid die forging aluminum alloy

Aluminium alloy		Mechanical properties	
Category	Grade	σ_b (MPa)	δ (%)
Al-Cu alloy	ZL201	458	16.7
Al-Cu-Mg alloy	2A14	506	5.8
	2A12	438	8.0
Al-Mg-Si-Cu alloy	2B50	380	13.0
Al-Zn-Mg-Cu alloy	7A04	563	5.5

b) High productivity

Liquid die forging is simple and maneuverable. The mechanized production and automated production are easy to organize for liquid die forging. In addition, the liquid die forging realizes the riserless casting and its process yield is as high as 90%~100%. Therefore, for the same smelting capacity, the output can be increased by 20%~30%.

c) High material utilization ratio

Liquid die forging eliminates the riser in conventional casting and the utilization ratio of metal melt can reach more than 90%. In addition, due to the use of high-strength metal mold, the liquid metal is closely attached to the cavity wall under pressure, the liquid forged workpiece has lower surface roughness and higher dimensional accuracy which is similar to that of die-casting: 0.25 mm in 100 mm to 0.6 mm in 500 mm. The machining allowance of casts is small, usually ±0.05 mm for the tolerance of nonferrous alloys. In addition, the scraps can be simply recycled. Thus the material consumption can be reduced.

d) High adaptability

In the solidification process of liquid die forging, each part is in the state of compressive stress, which is beneficial to the feeding of the cast and can prevent the casting cracks. Therefore, the liquid die forging process is not limited by the casting properties or plastic forming properties of alloys. It can be used for both casting alloys and deformation alloys, including non ferrous alloys, ferrous metals, composite materials, cast stones, etc.

The liquid die forging process is also not limited by the structure, shape and size of workpiece. The liquid die forging workpiece can have a wide wall thickness ranging from a few millimeters to several tens of millimeters. Even the complicated parts with different wall thickness can be obtained through a reasonable design process. Also the weight range is wide, ranging usually from some 10 g up to 5 kg; heavier casts can also be easily obtained.

In addition, liquid die forging can be used to produce metal matrix composites (MMCs); both continuous and discontinuous reinforcements with up to 45 vol.% of SiC have been produced.

e) Low energy consumption

Compared to forging, complex shaped parts can be produced in a single step by liquid die forging, as well as providing the possibility to produce undercuts and holes using retractable metal cores. The pressure applied is lower than that during forging but higher than that of in die-casting, ranging between 50 and 300 MPa, but typically between 50 and 140 MPa. Moreover, the process yield of liquid die forging is high. The energy for treating the waste products is greatly reduced. In addition, due to the small machining allowance, the energy used for machining is saved. Therefore the liquid die forging has low energy consumption.

f) Environment friendly

Liquid die forging does not require sand mold and does not generate the dust pollution. Furthermore, liquid die forging does not discharge waste liquid and harmful gas. Therefore, liquid die forging is a green manufacturing technology.

2) Disadvantages

Liquid die forging can result in the formation of macrosegregation when very high temperature gradients are generated by the pressurization of the alloy against the die surface, resulting in nonuniform microstructures and mechanical properties being obtained. Mainly when strongly segregating elements are present (i.e. Cu in Al alloys), the eutectic can also be squeezed out of the solid skeleton, exuding onto the casting surface, creating defects. Obviously, common foundry defects such as metallization, porosity, cold shuts, oxides, inclusions, etc. can be present if process parameters (pressure, lubrication, temperatures, alloy cleanliness) are not properly controlled.

In addition, cycle time for liquid die forging is longer than that of conventional die-casting, because of the slower injection speed and longer solidification times. Liquid die forging dies need thicker gates to inhibit premature solidification and to keep the flow speed as low as possible; as a consequence, liquid die forging parts have additional costs.

(4) Scope of application of liquid die forging

In principle, the liquid die forging technology is suitable for the production of parts with various shapes from various materials. Nevertheless, the liquid die forging technology can play its role in the following aspects.

1) In material aspect

Liquid die forging technology integrates the advantages of casting and forging. It can perfectly realize the forming of materials with poor casting and forging properties, such as alloy steel, hypereutectic alloys, composite, etc.

2) Shape and structure of workpiece

For the workpiece with complex shape and structure which are not suitable for casting and forging, such as air-tightness, water-tightness and complex curved surface parts, the liquid die forging technology can easily actualize the processing of these parts.

It can be seen that the application scope of liquid die forging is much wider than that of casting and forging.

(5) Classification, parameters and equipment of liquid die forging

1) Classification according to metal flow mode

According to the metal flow mode, liquid die forging can be divided into direct liquid forging, indirect liquid forging, etc.

a) Direct liquid forging

Direct liquid casting resembles the forging process and it is also called "liquid metal forging". The metal is transferred from the furnace to the die by a ladle or heated in a dosing furnace and then poured into the lower half die, which has a horizontal parting line [shown in Figure 7.25(a)]. When the appropriate amount of molten alloy has been poured, the upper die half moves downward and close on the lower die, acting as a punch [shown in Figure 7.25(b)]. In this way, the upper half die pushes the liquid metal in the cavity, forming the casting shape, and it applies a high pressure to the alloy melt until the solidification is complete.

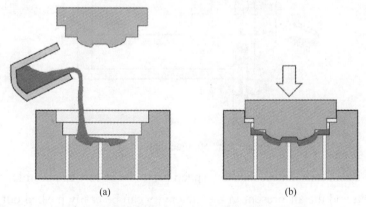

Figure 7.25 Direct liquid forging

An important advantage of this process is the lack of runner and gating systems as well as risers (biscuit or overflows), so that almost no scrap is produced and the yield is nearly 100%. The result is the reduced production times and costs; however, if fluctuations in ladling volume exceeding tolerances occur, then the scrap rate is high, as this volume directly affects the final dimension of the cast part.

Direct liquid forging guarantees castings free from porosity. On the other hand, since the use of runners helps in trapping inclusions before they enter the die, direct liquid forging suffers from oxides and inclusions. This technique does not have wide application in the market, except for relatively simple geometries.

The direct liquid forging can be divided into hydrostatic forging, extrusion hydraulic forging, etc.

b) Indirect liquid forging

Indirect liquid forging is mainly a vertical indirect liquid forging process. It combines the low-pressure die-casting (LPDC) bottom-up filling with high pressure die-casting (HPDC) injection.

The liquid metal is taken from the holding furnace by a ladle or poured through a dosing furnace into a shot sleeve placed slightly tilted below the die [shown in Figure 7.26(a)]. The shot sleeve is then moved in a vertical position and the piston, lifting from the bottom upwards, pushes the metal into the cavity by a sort of counter gravity process and holds the pressure until complete solidification of the alloy [shown in Figure 7.26(b)]. The machine is almost the same as an HPDC one; the key differences are the vertical shot sleeve and the tilt-docking injection unit. Compared to a direct liquid forging machine, it is more complex and expensive.

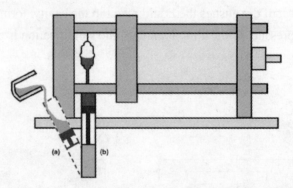

Figure 7.26　Indirect liquid forging

In order to avoid turbulence, the flow speed has to be as low as possible, so that the flow front remains flat and the air present in the die cavity can be easily pushed out through vents. The in-gate speed is about 1/100 that used in HPDC, as a direct result of lower velocity and larger gating areas. In particular, the critical velocity at the in-gate was demonstrated to be 0.5 m/s, significantly smaller than that of HPDC (30~50 m/s). Very seldom is vacuum needed.

Temperature and pressure are strictly controlled during the process; in particular, the pressure is constant during and after freezing, allowing the use of cores.

Some foundries place metallic mesh filters in front of the in-gate, as in LPDC, in order to reduce the number of inclusions.

Indirect liquid forging also offers the possibility of infiltrating preforms and producing composites.

The material yield is lower than that in the case of direct liquid forging (about 50%), with subsequent costs for scrap separation and recycling, but compared to direct liquid forging the use of a highly accurate metering system is not required.

Notwithstanding the more complex equipment necessary, the indirect liquid forging process experiences higher commercial use than the direct process.

2) Classification according to punch shape

According to the shape of punch, liquid die forging can be divided into flat punch liquid die forging, concave punch liquid die forging, convex punch liquid die forging and combined punch liquid die forging.

3) Process parameters

A number of parameters can affect the casting quality for both direct and indirect liquid forging:

a) The alloy and its quality

The first is the alloy and its quality; its melting temperature and thermal conductivity, in fact, affect die life and dictate the selection of casting parameters such as die temperature. Thus, liquid forging is preferentially employed for low melting temperature alloys such as Al and Mg. Also metal cleanliness is important to avoid inclusions of dross and oxides into the casting.

b) Melt quantity

Melt quantity is fundamental in direct liquid forging, as already mentioned, and precision control systems are needed to ensure dimensional control of casting. The die cavity can be designed taking into account the presence of an oversized appendix, placed in a noncritical area, to distribute any excess metal. Alternatively, Lynch proposed a compensating hydraulic piston and cylinder to control the exact quantity of metal in the die. Overflows can also be used.

c) Operating temperature

Operating temperature of the die cavity and punch has to be monitored as it affects the heat transfer rate and alloy cooling. In fact, too low temperature may result in premature solidification and cold laps in the casting, while too high die temperature can cause surface defects and metallization (casting and die welding). The die temperature usually ranges between 200 and 300 ℃ for Al and Mg alloys, the lower temperature being suitable for thicker section parts. A lubricant agent, typically graphite-based, should be used.

d) Time delay

Time delay is the interval between the actual pouring of liquid metal and the instant punch starts pushing the alloy into the die cavity. In order to reduce shrinkage porosity, a time delay is considered to allow cooling of the metal pool before squeezing. This time varies depending on melt/pouring temperature and complexity of the casting. Typically, a time delay of 6 s is used; for large aluminium components, this delay time could reach 1 min. According to some authors, the optimum time seems to be reached when the metal is midway between liquidus and solidus, that is, when a solid phase skeleton has formed in a twophase alloy and the metal presents a near zero fluidity; others recommend that the metal should be mainly liquid when the pressure is applied.

e) Magnitude and duration of applied pressure

Magnitude and duration of applied pressure are two other significant parameters. As the pressure influences the solidification temperature, it directly affects the microstructure and the mechanical properties of squeeze cast components. This may be explained by considering the Clausius-Clapeyron equation:

$$\frac{\Delta T}{\Delta P} = \frac{T_m \cdot \Delta V}{\Delta H_f} \qquad (7.11)$$

where T_m is the equilibrium melting temperature, ΔV is the difference between the specific volumes of liquid and solid, and ΔH_f is the latent heat of fusion. During solidification, both ΔV and ΔH_f are negative, because of shrinkage and heat released by the melt. Therefore, $\frac{\Delta T}{\Delta P}$ is positive and the increase in applied pressure induces a higher solidification temperature. The applied pressure also modifies the phase diagram by shifting the liquidus and solidus lines as well as the eutectic composition.

Additionally, the held pressure increases the heat transfer at the casting/die interface. In fact, during liquid forging the metal is forced against the die walls before any detachments occur, leading to very fast heat transfer and cooling rates. Pressures in the range of 50~140 MPa are typically used. Pressure duration usually varies from 30 to 120 s, depending on casting weight and geometry. However, according to a rule of thumb, 1 s/mm of section thickness of casting can be considered.

f) Die coating and lubrication agents

Die coating and lubrication agents depend on die materials and casting alloys; as already demonstrated, some release agents used in HPDC work well also in the case of liquid forging. Therefore, for nonferrous applications, common water-based colloidal graphite, sprayed onto the die and the plunger at the end of each cycle, has proved satisfactory. Some customized coatings specific for liquid forging are available on the market. To prevent contamination of the casts surface resulting from coating stripping from the die walls, the thickness of the coating/lubricant is limited to under 50 mm.

4) Equipment

A variety of liquid forging machines are now in use, both designed by researchers and manufactured by companies for mass production.

The direct liquid forging machines are simple and consist of a hydraulic press able to move a half die acting as a punch on the liquid alloy, guarantying the appropriate applied pressure and closing force.

Indirect machines are more complex and, according to the parting line orientation and the way the liquid metal is introduced into the cavity, they can be distinguished as vertical die closing and injection (VSC), horizontal die closing and injection, horizontal die closing and vertical injection (HVSC), and vertical die closing and horizontal injection machines.

The HVSC machine is the most widely used and its typical features include:
- Horizontal die clamping, with force ranging between 250 and 3,500 tons; it allows the production of castings both symmetrically and nonsymmetrically gated, increasing the number of potential parts to be cast;
- Vertical shot sleeve with tilt-docking and high-pressure delivery, which helps in maintaining molten metal temperature and minimizing turbulence, avoiding porosity formation and oxide skins.

With respect to the die, material, manufacturing process, heat treatment, and maintenance practice represent crucial aspects for the mass production of sound castings, as well as a very good die surface.

Liquid die forging undergo severe thermal and mechanical cyclic loadings that cause thermomechanical fatigue, erosion, and hot corrosion. Therefore, die steels used should provide:
- A high degree of cleanliness;
- Uniform microstructure;
- Chemical attack and thermal shock resistance;
- High resistance to mechanical and thermal fatigue;
- High ductility;
- Good hot hardness (thermochemical treatments);
- High thermal conductivity;
- High temper resistance;
- Adequate toughness;
- Good machinability;

7.4.2 Semisolid Forming

Semisolid forming

Semisolid forming of metals is a fascinating technology offering the opportunity to manufacture net-shaped metal components of complex geometry in a single forming operation. At the same time, high mechanical properties can be achieved due to the unique microstructure and flow behaviour. Successful semi-solid forming processes require narrow tolerances in all process steps, including feedstock generation, reheating and the forming process. It is this strong and highly nonlinear interrelation between the parameters of each process step on the one hand and the material microstructure and flow behaviour on the other which still causes a challenge for scientific understanding and economic mass production.

(1) Overview of semisolid forming

1) Basic conception

The semisolid slurry is used for the subsequent processing such as casting, forging, rolling, etc. The processing method which is neither complete liquid forming nor complete solid forming is called semisolid metal forming[3]. The evolution of dendrites to spherical particles in semisolid slurry is shown in Figure 7.27. The semisolid cast structure which is quite different from that forming in the conventional casting is shown in Figure 7.28. The cast structure of conventional casting appears to be a traditional dendritic structure. Nevertheless, the cast structure of semisolid casting appears to be a granular structure resulting from the spherical primary solid particles.

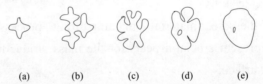

(a)　　(b)　　(c)　　(d)　　(e)

Figure 7.27　Evolution of dendrites to spherical particles

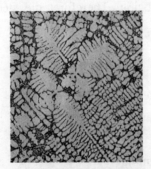 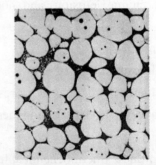

(a) Cast structure of conventional casting　　(b) Cast structure of semisolid casting

Figure 7.28　Cast structure forming in different casting manners

2) Semisolid metal

Semisolid metal refers to the metal in the solid-liquid two-phase region, including semisolid slurry and semisolid billet. The semisolid metal has excellent rheology and thixotropy.

a) Rheology

Rheology refers to the deformation and flow properties of a semisolid slurry under external force. Although the viscosity of semisolid slurry increases with the volume fraction of solid component during stirring, a good fluidity of semisolid slurry may be still maintained even when the volume fraction of solid component increases up to 40%. For example, the macroscopic fluidity of most alloys will disappear generally when the solid fraction reaches to 20%~30%. However, some alloys such as aluminum alloy still have some fluidity due to the dispersion of granular solid components even if the solid fraction reaches up to 40% during stirring.

b) Thixotropy

Thixotropy refers to the deformation and flow properties of a semisolid billet under external force. The semisolid billet has obvious superplastic effect and filling ability in forming. Furthermore, its deformation resistance is also small and hence it can be deformed at a higher speed.

3) Measurement of steady-state rheological behavior

An absolute viscometer is an instrument in which the measurement of viscosity can be traced back to the test results in the fundamental physical units of force in Newtons, dimensions of the sensor system in meters, and time interval in seconds. Viscosity is then defined in the unit of Pascal seconds (Pa·s). The requirements for absolute viscometry are rather difficult to achieve for the following reasons:

- The samples must be tested in a way that lends itself to rigorous mathematical evaluation, in terms of the calculation of shear stresses and shear rates at the location of interest.
- The test conditions must be chosen to take into account the conditions assumed in the valuation of viscosity (i.e. there must be laminar flow, the sample must be homogeneous and not undergo any physical or chemical change, and there must be no elasticity). It is not possible to achieve this for thixotropic materials as they do undergo changes during testing. All tests are therefore an approximation to the required conditions.

The advantage of absolute viscometry is that the results are independent of the type or design of viscometer used. There are three important types commercially: falling-ball rheometers, capillary viscometers, and rotational viscometers. Falling-ball rheometers are not suitable for semisolid metals because a transparent fluid is required and the mathematical treatment is not applicable to non-Newtonian fluids. Capillary viscometers measure the resistance of a liquid to flow through a capillary tube. Two pressure transducers measure the pressure drop over the length of the capillary. The viscosity is then calculated from this pressure drop and the flow rate. For high-shear-rate experiments, the advantage is that any heat generated as a result of the high shear rate is carried out of the end of the capillary. This minimizes any errors in viscosity caused by changes in sample temperature as a result of this heat. However, the time for which the sample is sheared is limited by the shear rate, making these viscometers unsuitable for testing thixotropy as the sample cannot be sheared for extended periods of time. Rotational viscometers are therefore the most common method of measuring apparent viscosity.

4) Classification of semisolid forming

Semisolid forming can be divided into rheoforming and thixoforming (shown in Figure 7.29).

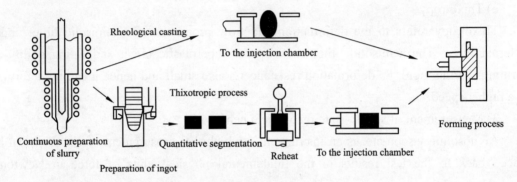

Figure 7.29 Rheoforming and thixoforming

a) Rheoforming

Rheoforming refers to the forming of semisolid slurry. Generally, all the conventional processing manners can be used for semisolid forming. However, there are some problems in the preparation of semisolid slurry and this influences the development of rheoforming. So far, the available rheoforming technology mainly includes casting, rolling, extrusion, etc.

b) Thixoforming

Thixoforming refers to the forming of semisolid billet. The semisolid billet is obtained by reheating the block to the solid-liquid two-phase region, so conventional processing methods can be used for the thixoforming. This forming method is a complement to rheoforming. The typical thixoforming technology mainly includes forging, rolling, extrusion etc.

(2) Characteristics of semisolid forming

Semisolid forming is such a new forming method that combines the advantages of liquid forming and solid forming. Its processing temperature is lower than that of liquid forming and its deformation resistance is smaller than that of solid forming. Therefore, it can realize the processing of complex-shaped parts with large deformation and is considered as the most promising materials forming method in the 21st century. The characteristics of semisolid forming are as follows:

- Wide range of application;
- Long service life of mold;
- Rapid speed of forming;
- Low cost and energy consumption;
- High quality of workpiece.

(3) Preparation of semisolid slurry

Since the 1970s, researchers have conducted systematic researches on the preparation of semisolid slurry of aluminum alloys, magnesium alloys, copper alloys, etc. The available preparation methods of semisolid slurry can be divided into mechanical stirring method, electromagnetic stirring method, electromagnetic mechanical stirring method, ultrasonic vibration method, cooling slope method, etc.

1) Mechanical stirring method

Mechanical stirring method is the earliest used method which was developed by Flemings of MIT in the early 1970s. Its working principle is as follows. During the solidification process of metal melt, the vigorous stirring is applied by stirring rod and the primary solid phase dendrites are fully broken to pieces, resulting in a solid-liquid coexisting slurry which the spherical primary solid particles uniformly suspends in the remaining liquid (shown in Figure 7.30). In the mechanical stirring, the heat loss of the mechanical stirring component at high temperature is large, that is, the pollution to the semisolid slurry is serious. Therefore, the stirring component should have an excellent high-temperature performance.

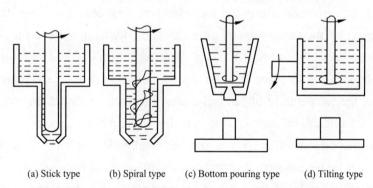

(a) Stick type (b) Spiral type (c) Bottom pouring type (d) Tilting type

Figure 7.30　Schematic diagram of mechanical stirring devices

2) Electromagnetic stirring method

The working principle of electromagnetic stirring method is as follows: During the solidification process of metal melt, the vigorous stirring is applied by the electromagnetic induction force and the primary solid phase dendrites are fully broken to pieces, resulting in a solid-liquid coexisting slurry which the spherical primary solid particles uniformly suspends in the remaining liquid (shown in Figure 7.31). The electromagnetic stirring effect can be influenced by stirring power, stirring time, cooling rate, metal melt temperature, pouring speed, etc. In the electromagnetic stirring, the gas entrapment and the pollution to the semisolid slurry are slight. The semisolid slurry can be continuously produced and its output can be large and thus the electromagnetic stirring method is the most widely used in the industry, especially in the production of aluminum alloy slurry.

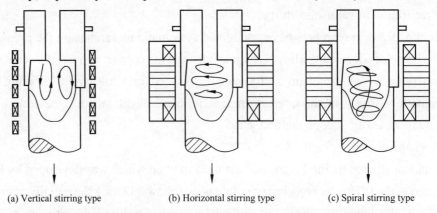

(a) Vertical stirring type　　(b) Horizontal stirring type　　(c) Spiral stirring type

Figure 7.31　Schematic diagram of electromagnetic stirring devices

3) Electromagnetic mechanical stirring method

Electromagnetic mechanical stirring method is a newly developed preparation method of semisolid slurry. Its stirring device is shown in Figure 7.32. The mechanical stirrer is controlled by the motor and can move up and down in the vertical direction, resulting in the large-scale flow of semisolid slurry in the whole mixing chamber. The electromagnetic stirring is achieved by the three pairs of poles (N and S) which can exert a circumferential electromagnetic force to the semisolid slurry. The cooling holes and heating holes on the graphite crucible are used to place the cooling tube and the heating rod which are used to precisely control the temperature of semisolid slurry together with the thermocouple. Argon gas protection is applied throughout the stirring to prevent the semisolid slurry from oxidizing. This method can effectively avoid the segregation of metal components during stirring and can ensure the uniform distribution of primary solid phase particles. Therefore, it has obvious advantages and wide applications in preparing semisolid slurry of composite.

4) Ultrasonic vibration method

Ultrasonic vibration method is shown in Figure 7.33. Its basic principle of preparing semisolid slurry is as follows: The ultrasonic mechanical vibration wave is used to disturb the solidification process of metal melt aiming at refining the grains and obtaining the non-dendritic slurry. There are two ways for applying ultrasonic vibration waves to metal melts. One way is that the vibrator applies the ultrasonic vibration wave directly to the metal melt. The other way is that the vibrator applies the ultrasonic vibration wave to the mold and then the mold applies the ultrasonic vibration to the metal melt. Tests have shown that ultrasonic vibration can help the alloy liquid to obtain a semisolid slurry with small spherical primary solid particles.

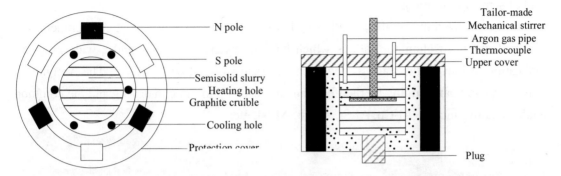

Figure 7.32 Electromagnetic mechanical stirring device

5) Cooling slope method

Cooling slope method is such a method that operates as follows: The melt first flows through the cooling slope. After partial cooling and strong trundling, it flows into the mold. By controlling the temperature of mold, a semisolid slurry with the required solid phase volume fraction can be obtained (shown in Figure 7.34).

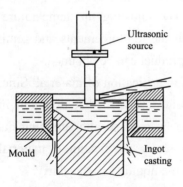

Figure 7.33 Schematic diagram of ultrasonic vibration method

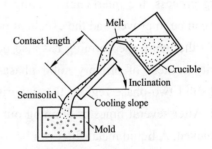

Figure 7.34 Schematic diagram of cooling slope method

6) Other methods

In addition to the above methods, several available preparation methods such as turbulence effect method, near liquid phase line solidification method, sputtering deposition method, strain-induced melt activation method has been developed.

(4) Development of semisolid forming

1) Development of rheoforming

Compared with thixoforming, rheoforming eliminates the secondary heating and realizes the processing of part directly from semisolid slurry. Thus its production process is short and the cost is relatively low. In recent years, many researchers have paid much attention to this technology. Thus rheoforming has made rapid progress and becomes a new hot spot in the field of semisolid forming.

In the rheoforming of alloys with a wide range of liquid-solid coexistence temperature, the rheoforming of aluminum alloys which have a low melting point and a wide range of application achieves the fastest development (shown in Figure 7.35). The weight of aluminum alloy part forming by rheoforming can reach up to more than 7 kg and the common available alloys include Al-Cu, Al-Si, Al-Pb, Al-Ni, etc.

(a) Crankshaft　　　　　　　　　(b) Connecting rod

Figure 7.35　Semisolid formed aluminum alloy parts

In addition, rheoforming has been applied to the preparation of composite materials. Since the semisolid slurry has good viscosity and fluidity in the liquid-solid two-phase region, the non-metallic reinforcements can be added easily. By adjusting the temperature and stirring process, the interfacial bonding strength between the reinforcements and semisolid slurry can be improved and the excellent performance of product can be obtained.

Furthermore, rheoforming has also been used for the purification of material. Since the impurity content of primary solid phase particles is quite less than that of liquid in the liquid-solid two-phase region, the impurity in slurry can be reduced after draining out the liquid. After several times of draining out the liquid from slurry, purification of material can be achieved. A liquid metal filter can be used to drain out the liquid from the slurry.

2) Development of thixoforming

The technological process of thixoforming is shown in Figure 7.36.

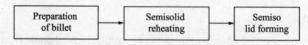

Figure 7.36　Technological process of thixoforming

The semisolid reheating of billet in an induction heating furnace is an important process which aims at obtain the required solid phase volume fraction to create a favorable condition for the subsequent forming. Even an error of 1~2 K can significantly affect the structure of billet, thus there is a strict requirement to the heating temperature and speed.

In recent years, the thixoforming of magnesium alloy has been greatly developed. The semisolid thixotropic injection moulding of magnesium alloy is shown in Figure 7.37.

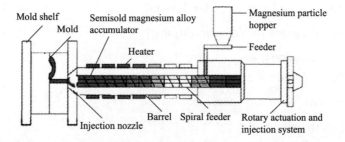

Figure 7.37　Semisolid thixotropic injection moulding of magnesium alloy

(5) Parts of semisolid forming in rail traffic

The magnesium alloy gearboxes produced semisolid thixotropic injection forming have been used in CRH2 EMUs (200 km/h) (shown in Figure 7.38). Compared with the hot chamber die casts, their deviation and solidification shrinkage are reduced by 72% and 44%, respectively. Their elongation is greatly increased, which exceeds the technical standards of the A291D alloy in the ASTM Specification. Compared with the high-performance aluminum alloy parts, their performance price ratio is significantly improved. These magnesium alloy oil pump boxes and gearboxes reduce the weight of high speed trains, meet the needs of rail traffic development and provide a good opportunity for the small and medium enterprises.

Figure 7.38　Magnesium alloy gearbox

New processing technologies such as liquid die forging and semisolid forming provide various preparation methods for new materials. With the further development of science and the invention of new processing technologies, the unprecedented materials which can meet the urgent needs of society will be produced.

Words and Expressions

rheological feeding [riːəˈlɒdʒɪkəl ˈfiːdɪŋ]	流变补缩
notwithstanding [ˌnɒtwɪθˈstændɪŋ] despite anything to the contrary (usually following a concession)	*adv.* 虽然

squeeze casting [skwiːz ˈkæstɪŋ]	挤压铸造
process chart [ˈprəʊses tʃɑːt]	工艺流程图
vacuum die-casting [ˈvækjuəm daɪ ˈkæstɪŋ]	真空模铸
production efficiency [prəˈdʌkʃən ɪˈfɪʃənsi]	生产效率
turbulence [ˈtɜːbjʊləns] unstable flow of a liquid or gas	n. 紊流
air entrapment [eə ɪnˈtræpmənt]	卷气
porosity [pɔːˈrɒsəti] the property of being porous; being able to absorb fluids	n. 多孔性
fine-grained [faɪn greɪnd]	细晶的
superb [suːˈpɜːb] of surpassing excellence	adj. 极好的
hydrostatic pressure [ˌhaɪdrəˈstætɪk ˈpreʃə]	静水压力
gating system [geɪtɪŋ ˈsɪstəm]	浇注系统
runner [ˈrʌnə] someone who imports or exports without paying duties	n. 浇道
fin line [fɪn laɪn]	鳍线
maneuverable [məˈnuːvərəbl] capable of maneuvering or changing position	adj. 操作方便的；移动熟练的
utilization ratio [ˌjuːtɪlaɪˈzeɪʃən ˈreɪʃɪəʊ]	利用率
process yield [ˈprəʊses jiːld]	工艺出品率
machining allowance [məˈʃiːnɪŋ əˈlaʊəns]	加工余量
environment friendly [ɪnˈvaɪrənmənt ˈfrendli]	环境友好
macrosegregation [mækrəʊˌsegrɪˈgeɪʃən]	宏观偏析
in-gate [ˈɪngeɪt]	内浇口
extrusion hydraulic forging [ɪkˈstruːʒən haɪˈdrɔːlɪk ˈfɔːdʒɪŋ]	挤压静压锻造
coupler knuckle [ˈkʌplə ˈnʌkl]	钩舌
feedstock [ˈfiːdstɒk] the raw material that is required for some industrial process	n. 原料
spherical particle [ˈsferɪkl ˈpɑːtɪkl]	球形颗粒
viscometer [vɪˈskɒmɪtə]	n. 黏度计

a measuring instrument for measuring viscosity	
laminar flow [ˈlæmɪnə fləʊ]	层流
thixoforming [θɪkˈsɒfɔːmɪŋ] shear thinning shaping	n. 触变成形
rheology [rɪˈɒlədʒi] the branch of physics that studies the deformation and flow of matter	n. 流变性
thixotropy [θɪkˈsɒtrəpi] shear thinning property	n. 触变性
electromagnetic [ɪˌlektrəʊmægˈnetɪk] pertaining to or exhibiting magnetism produced by electric charge in motion	adj. 电磁的
deviation [diːvɪˈeɪʃən] a variation that deviates from the standard or norm	n. 偏差

Notes

[1] 本句为让步状语从句，全句可译为：虽然有着百年的历史，但是液态模锻在 1960 年后才在北美、欧洲和日本实现商业化，用于生产铝制汽车零部件。

[2] 本句为过去分词作后置定语句型，全句可译为：在合金凝固之前、过程中和之后施加的压力保证了金属和模具壁之间的接触，增加了热流速，因此提高了冷却速率，从而产生了细晶组织。

[3] 本句为 neither … nor 句型，全句可译为：既不是完全液态也不是完全固态成形的加工方法称为半固态金属成形。

Questions for Discussion

1. What is the liquid die forging?
2. What is the current problem of liquid die forging?
3. What are the main advantages of liquid die forging?
4. When will the common foundry defects present in the liquid die forging?
5. Why is the application scope of liquid die forging much wider than that of casting and forging?
6. What characteristics does the semisolid metal possess?
7. What characteristics does the semisolid forming possess?
8. How many methods are there in the preparation of semisolid slurry? What are they?

References

[1] 雷源忠. 我国机械工程研究进展与展望[J]. 机械工程学报, 2009, 45(5): 1-11.

[2] 唐一平. 机械工程专业英语[M]. 3 版. 北京：电子工业出版社, 2017.

[3] 施平. 机械工程专业英语[M]. 哈尔滨：哈尔滨工业大学出版社, 2017.

[4] 马玉录, 刘东学. 机械设计制造及其自动化专业英语[M]. 3 版. 北京：化学工业出版社, 2017

[5] 王群, 叶久新. 模具专业英语:设计·制造·报价·结算[M]. 北京：机械工业出版社, 2008.

[6] 唐一平. 先进制造技术:英文版 [M]. 4 版. 北京：科学出版社, 2017.

[7] 王晓江. 机械制造专业英语[M]. 2 版. 北京：机械工业出版社, 2017.

[8] 彭伟, 肖帆. 先进制造技术:英文版 [M]. 北京：清华大学出版社, 2013.

[9] SINGLETON R, MARSHALL M B, LEWIS R, et al. Rail grinding for the 21st century: Taking a lead from the aerospace industry[J]. Proceedings of the Institution of Mechanical Engineers Part F Journal of Rail & Rapid Transit, 2015(5), 229.

[10] YANG Y, QIU W S, ZENG W, et al. A prediction method of rail grinding profile using non-uniform rational B-spline curves and Kriging model[J]. 中南大学学报（英文版）, 2018, 25(1):230-240.

[11] KUFFA M, ZIEGLER D, PETER T, et al. A new grinding strategy to improve the acoustic properties of railway tracks[J]. Proceedings of the Institution of Mechanical Engineers Part F: Journal of Rail & Rapid Transit, 2018(1), 232.

[12] ASTROM, KARL J. 自适应控制[M]. 北京：科学出版社, 2003.

[13] NISE N S .Control systems engineering [M]. 6th. New Jersey：Wiley, 2011.

[14] MCMILLAN G K. Industrial applications of PID control[J]. Advances in Industrial Control, 2012:415-461.

[15] HARASHIMA F, TOMIZUKA M, FUKUDA T. Mechatronics—"What Is It, Why, and How?" [J]. IEEE/ASME Transactions on Mechatronics, 1996, 1(1): 1-4.

[16] SCHROEDER B A. On-line monitoring: A tutorial[J]. Computer, 1995 (6): 72-78.

[17] BOURGEOIS W, BURGESS J E, STUETZ R M. On-line monitoring of wastewater quality: a review[J]. Journal of Chemical Technology & Biotechnology: International Research in Process, Environmental & Clean Technology, 2001, 76(4): 337-348.

[18] BLODT M, BONACCI D, REGNIER J, et al. On-line monitoring of mechanical faults in variable-speed induction motor drives using the Wigner distribution[J]. IEEE Transactions on Industrial Electronics, 2008, 55(2): 522-533.

[19] NANDI S, TOLIYAT H A, LI X. Condition monitoring and fault diagnosis of electrical motors: a review[J]. IEEE transactions on energy conversion, 2005, 20(4): 719-729.

[20] TEODORESCU P P. Mechanical systems, classical models. Volume II: Mechanics of discrete and continuous systems[M]. New York: Springer, 2009.

[21] MARWEDEL P. Embedded system design[M]. New York: Springer, 2006.

[22] SANGIOVANNI-VINCENTELLI A, NATALE M D. Embedded System Design for Automotive Applications[J]. Computer, 2007, 40(10): 42-51.

[23] GAJSKI D D, ABDI S, GERSTLAUER A, et al. Embedded system design: modeling, synthesis and verification[M]. Berlin: Springer Science & Business Media, 2009.

[24] SGROI M, LAVAGNO L, SANGIOVANNI-VINCENTELLI A. Formal models for embedded system design[J]. IEEE Design & Test of Computers, 2000, 17(2): 14-27.

[25] POLANSKY L, ROSENBOOM D, BURK P. HMSL: Overview (version 3.1) and notes on intelligent instrument design[C]//ICMC. 1987.

[26] LIN Y, JI H. Intelligent instrument and its development [J]. Instrumentation Technology, 2003, 1: 37-39.

[27] BHUYAN M. Intelligent instrumentation: principles and applications[M]. Boca Raton: CRC Press, 2010.

[28] KNUTH K H, ERNER P M, FRASSO S. Designing intelligent instruments[C]//AIP Conference Proceedings. AIP, 2007, 954(1): 203-211.

[29] BIROLINI A. Reliability engineering: theory and practice[M]. 7th ed. Heidelberg: Springer-Verlag, 2014.

[30] CHOI S K, CANFIELD R A, GRANDHI R V. Reliability-based structural design[M]. London: Springer-Verlag, 2007.

[31] IWNICKI S. Handbook of railway vehicle dynamics[M]. Boca Raton: CRC Press, 2006.

[32] BUNI S, GOODALL R, MEI T X, et al. Control and monitoring for railway vehicle dynamics[J]. Vehicle System Dynamics, 2007, 45(7-8): 743-779.

[33] LAKUŠIĆ S, AHAC M. Rail traffic noise and vibration mitigation measures in urban areas[J]. Technical Gazette, 2012, 19(2): 427-435.

[34] THOMPSON D. Railway noise and vibration: mechanisms, modelling and means of control[M]. Oxford: Elsevier, 2010.

[35] WANG X X, HU G Z, XIAO S N, et al. Development and application of CAD technology in railway vehicle studying[J]. Advanced Materials Research, 2014, 1037: 468-473.

[36] WILLIAM D C. Materials science and engineering: an introduction[M]. New Jersey: Wiley, 1985.

[37] MITCHELL B S. An Introduction to materials engineering and science: for chemical and materials engineers[M]. New Jersey: Wiley, 2005.

[38] HIRT G, KOPP R. Thixoforming: semi-solid metal processing[M]. New Jersey: Wiley, 2009.

[39] SHIUE R K, WU S K, HUNG C M. Infrared repair brazing of 403 stainless steel with a nickel-based braze alloy[J]. Metallurgical & Materials Transactions A, 2002, 33(6):1765-1773.

[40] PATEL H A, RASHIDI N, CHEN D L, et al. Cyclic deformation behavior of a super-vacuum die cast magnesium alloy[J]. Materials Science & Engineering A, 2012, 546(3):72-81.

[41] FOROOZMEHR A, KERMANPUR A, ASHRAFIZADEH F, et al. Investigating microstructural evolution during homogenization of the equiatomic NiTi shape memory alloy produced by vacuum arc remelting[J]. Materials Science & Engineering A, 2011, 528(27):7952-7955.

[42] NAYAN N, GOVIND, SAIKRISHNA C N, et al. Vacuum induction melting of NiTi shape memory alloys in graphite crucible[J]. Materials Science & Engineering A, 2007, 465(1):44-48.

[43] MEI L C, PRICE A, BELLOWS R S. Advanced solidification processing of an industrial gas turbine engine component[J]. JOM, 2003, 55(3):27-31.

[44] CUI R, GAO M, ZHANG H, et al. Interactions between TiAl alloys and yttria refractory material in casting process[J]. Journal of Materials Processing Tech, 2010, 210(9):1190-1196.

[45] ZHENG L, XIAO C, ZHANG G, et al. Primary α phase and its effect on the impact ductility of a high Cr content cast Ni-base superalloy[J]. Journal of Alloys & Compounds, 2012, 527(1):176-183.

[46] CAMPBELL J, TIRYAKIOĞLU M. Bifilm defects in Ni-Based alloy castings[J]. Metallurgical & Materials Transactions B, 2012, 43(4):902-914.

[47] RASHID A K M B, CAMPBELL J. Oxide defects in a vacuum investment-cast ni-based turbine blade[J]. Metallurgical & Materials Transactions A, 2004, 35(7):2063-2071.

[48] LI D Z, CAMPBELL J, LI Y Y. Filling system for investment cast Ni-base turbine blades[J]. Journal of

Materials Processing Tech, 2004, 148(3):310-316.

[49] PANG H T, HOBBS R A, STONE H J, et al. Solution heat treatment optimization of fourth-generation single-crystal nickel-base superalloys[J]. Metallurgical & Materials Transactions A, 2012, 43(9):3264-3282.

[50] NAYAN N, GOVIND, NAIR K S, et al. Studies on Al–Cu–Li–Mg–Ag–Zr alloy processed through vacuum induction melting (VIM) technique[J]. Materials Science & Engineering A, 2007, 454(16):500-507.

[51] TSAI D C, HWANG W S. Numerical simulation of the solidification processes of copper during vacuum continuous casting[J]. Journal of Crystal Growth, 2012, 343(1):45-54.

[52] SOINILA E, PIHLAJAMÄKI T, BOSSUYT S, et al. A combined arc-melting and tilt-casting furnace for the manufacture of high-purity bulk metallic glass materials[J]. Review of Scientific Instruments, 2011, 82(7):651.

[53] HOSKO J, JANOTOVA I, SVEC P, et al. Preparation of thin ribbon and bulk glassy alloys in CoFeBSiNb(Ga) using planar flow casting and suction casting methods[J]. Journal of Non-Crystalline Solids, 2012, 358(12-13):1545-1549.

[54] RAFFERTY A, BAKIR S, BRABAZON D, ET AL. Calibration and characterisation with a new laser-based magnetostriction measurement system[J]. Materials & Design, 2009, 30(5):1680-1684.

[55] CASTILLOOYAGÜE R, OSORIO R, Osorio E, et al. The effect of surface treatments on the microroughness of laser-sintered and vacuum-cast base metal alloys for dental prosthetic frameworks.[J]. Microscopy Research & Technique, 2012, 75(9):1206-1212.

[56] OYAGÜE R C, SÁNCHEZ-TURRIÓN A, LÓPEZ-LOZANO J F, et al. Vertical discrepancy and microleakage of laser-sintered and vacuum-cast implant-supported structures luted with different cement types[J]. Journal of Dentistry, 2012, 40(2):123-130.

[57] HUDA D. Metal-matrix composites: manufacturing aspects[J]. Journal of Materials Processing Technology, 1993, 400(8):279–282.

[58] Jr W C H. Commercial processing of metal matrix composites[J]. Materials Science & Engineering A, 1998, 244(1):75-79.

[59] SEO Y H, KANG C G. Effects of hot extrusion through a curved die on the mechanical properties of SiC p /Al composites fabricated by melt-stirring[J]. Composites Science & Technology, 1999, 59(5):643-654.

[60] SKOLIANOS S. Mechanical behavior of cast SiC p reinforced Al-4.5%Cu-1.5%Mg alloy[J]. Materials Science & Engineering A, 1996, 210(1–2):76-82.

[61] KANG C G, YOON J H, SEO Y H. The upsetting behavior of semi-solid aluminum material fabricated by a mechanical stirring proces[J]. Journal of Materials Processing Technology, 1997, 66(1–3):30-38.

[62] HANUMANTH G S, IRONS G A. Particle incorporation by melt stirring for the production of metal-matrix composites[J]. Journal of Materials Science, 1993, 28(9):2459-2465.

[63] LEE J C, BYUN J Y, OH C S, et al. Effect of various processing methods on the interfacial reactions in SiC p /2024 Al composites[J]. Acta Materialia, 1997, 45(12):5303-5315.

[64] LIM C S, CLEGG A J. The production and evaluation of metal-matrix composite castings produced by a pressure-assisted investment casting process[J]. Journal of Materials Processing Technology, 1997, 67(1–3):13-18.

[65] MICHAUD V, MORTENSEN A. Infiltration processing of fibre reinforced composites: governing phenomena[J]. Composites Part A Applied Science & Manufacturing, 2001, 32(8):981-996.

[66] SEO Y H, KANG C G. The effect of applied pressure on particle-dispersion characteristics and mechanical properties in melt-stirring squeeze-cast SiC p /Al composites[J]. Journal of Materials Processing Technology, 1995, 55(3):370-379.

[67] XU Q, HAYES R W, JR W H H, et al. Mechanical properties and fracture behavior of layered 6061/SiC p,

composites produced by spray atomization and co-deposition[J]. Acta Materialia, 1998, 47(1):43-53.

[68] GUPTA M, SURAPPA M K, QIN S. Effect of interfacial characteristics on the failure-mechanism mode of a SiC reinforced A1 based metal-matrix composite[J]. Journal of Materials Processing Technology, 1997, 67(1–3):94-99.

[69] LIN J T, BHATTACHARYYA D, LANE C. Machinability of a silicon carbide reinforced aluminium metal matrix composite[J]. Wear, 1995, s 181–183(95):883-888.

[70] SAMUEL A M, SAMUEL F H. Foundry Aspects of Particulate Reinforced Aluminum MMCs: Factors Controlling Composite Quality[J]. Key Engineering Materials, 1995, 104-107:65-98.

[71] LLOYD D J. Aspects of fracture in particulate reinforced metal matrix composites[J]. Acta Metallurgica Et Materialia, 1991, 39(1):59-71.

[72] EFTEKHARI A, TALIA J E, MAZUMDAR P K. Influence of surface condition on the fatigue of an aluminum-lithium alloy (2090-T3)[J]. Materials Science & Engineering A, 1995, 199(2):L3–L6.

[73] EJIOFOR J U, REDDY R G. Characterization of pressure-assisted sintered Al–Si composites[J]. Materials Science & Engineering A, 1999, 259(2):314-323.

[74] ROHATGI P K. Future directions in solidification of metal matrix composites[J]. Key Engineering Materials, 1995, 104-107(1):293-312.

[75] SURAPPA M K. Microstructure evolution during solidification of DRMMCs (Discontinuously reinforced metal matrix composites): State of art[J]. Journal of Materials Processing Technology, 1997, 63(1–3):325-333.

[76] BONOLLO F, GUERRIERO R, SENTIMENTI E, et al. The effect of quenching on the mechanical properties of powder metallurgically produced AlSiC (particles) metal matrix composites[J]. Materials Science & Engineering A, 1991, 144(1–2): 303-309.

[77] KRISHNAN B P, SURAPPA M K, ROHATGI P K. The UPAL process: a direct method of preparing cast aluminium alloy-graphite particle composites[J]. Journal of Materials Science, 1981, 16(5): 1209-1216.

[78] LIN C B, MA C L, CHUNG Y W. Composites for die casting[J]. Journal of Materials Processing Technology, 1998, 84(1-3): 236-246.

[79] CHARLES D. Metal matrix composites. Ready for take-off[J]. Metals & Materials Bury St Edmunds, 1990, 6(2):78-82.

[80] SUÃ©RY M, L'Esperance G. Interfacial Reactions and Mechanical Behaviour of Aluminium Matrix Composites Reinforced with Ceramic Particles[J]. Key Engineering Materials, 1993, 79-80:33-46.

[81] RALPH B, YUEN H C, LEE W B. The processing of metal matrix composites —an overview[J]. Journal of Materials Processing Technology, 1997, 63(1-3):339-353.

[82] ROBERT L M. Machine elements in mechanical design[M]. 3th ed. New Jersey: Prentice Hall. 1992.

[83] ROBERT O. PARMLEY P E., Machine devices and components[M]. New York: McGraw-Hill, 2004.

[84] ROBERT L. Norton, Design of machinery: an introduction to the synthesis and analysis of mechanisms and machines[M]. New York: McGraw-Hill, 2001.

[85] NEIL SCLATER, NICHOLAS P. Chironis, Mechanisms & mechanical devices sourcebook [M]. 3th ed. New York: McGraw-Hill, 2001.

[86] 叶邦彦, 陈统坚. 机械工程英语[M]. 2 版. 北京：机械工业出版社, 2009.

[87] STACEY M H. Production and characterisation of fibres for metal matrix composites[J]. Metal Science Journal, 2014, 4(3):227-230.

[88] HIKOSAKA T, IMAI T. Effect of hot rolling on superplasticity of a SiC/6061 aluminum alloy composite made by a vortex method[J]. Reports of Industrial Research Institute Aichi Prefectural Government, 1997,

36(2):145-150.

[89] GHOSH P K, RAY S. Fabrication and properties of compocast aluminium-alumina particulate composite[J]. Indian Journal of Chemical Technology, 1988, 26(2):83-94.

[90] LACOSTE E, MANTAUX O, DANIS M. Numerical simulation of metal matrix composites and polymer matrix composites processing by infiltration: a review[J]. Composites Part A, 2002, 33(12):1605-1614.

[91] SU H, GAO W, FENG Z, et al. Processing, microstructure and tensile properties of nano-sized Al 2 O 3, particle reinforced aluminum matrix composites[J]. Materials & Design, 2012, 36(Complete):590-596.

[92] MORTENSEN A, CORNIE J A, Flemings M C. Solidification processing of metal-matrix composites[J]. Metallurgical Reviews, 1988, 37(1):101-128.

[93] MOON H K, CORNIE J A, FLEMINGS M C. Rheological behavior of SiC particulate-(Al-6.5wt.%Si) composite slurries at temperatures above the liquidus and within the liquid + solid region of the matrix[J]. Materials Science & Engineering A, 1991, 144(1–2):253-265.

[94] FLEMINGS M C. Behavior of metal alloys in the semisolid state[J]. Metallurgical Transactions B, 1991, 22(3):269-293.

[95] SCUDINO S, LIU G, PRASHANTH K G, et al. Mechanical properties of Al-based metal matrix composites reinforced with Zr-based glassy particles produced by powder metallurgy[J]. Acta Materialia, 2009, 57(6): 2029-2039.

[96] ABBASIPOUR B, NIROUMAND B, VAGHEFI S M M. Compocasting of A356-CNT composite[J]. Transactions of Nonferrous Metals Society of China, 2010, 20(9): 1561-1566.

[97] CESCHINI L, MINAK G, MORRI A. Tensile and fatigue properties of the AA6061/20 vol% AlO and AA7005/10 vol% AlO composites[J]. Composites Science & Technology, 2006, 66(2):333-342.

[98] NAHER S, BRABAZON D, Looney L. Development and assessment of a new quick quench stir caster design for the production of metal matrix composites[J]. Journal of Materials Processing Tech, 2005, 166(3):430-439.

[99] TEKMEN C, OZDEMIR I, COCEN U, et al. The mechanical response of Al–Si–Mg/SiC p, composite: influence of porosity[J]. Materials Science & Engineering A, 2003, 360(1-2):365-371.

[100] FU J, WANG K. Modelling and Simulation of Die Casting Process for A356 Semi-solid Alloy[J]. Procedia Engineering, 2014, 81:1565-1570.

[101] BUI R T, OUELLET R, KOCAEFE D. A two-phase flow model of the stirring of Al-SiC composite melt[J]. Metallurgical & Materials Transactions B, 1994, 25(4):607-618.

[102] KOCAEFE D, BUI R T. A one-phase model of the mixing of Al-SiC composite melt[J]. Metallurgical & Materials Transactions B, 1996, 27(6):1015-1023.

[103] MADA M, AJERSCH F. Rheological model of semi-solid A356-SiC composite alloys. Part I: Dissociation of agglomerate structures during shear[J]. Materials Science & Engineering A, 1996, 212(1):157-170.

[104] FOAKES R A. Simulation of the stir casting process[J]. Journal of Materials Processing Tech, 2003, 143(1):567-571.

[105] WESTGARD J V S. Mechanical behavior of materials[M]. New York: McGraw-Hill, 1990.

[106] CRAIG B D, ANDERSON D S, International A. Handbook of corrosion data[M]. Almere: ASM International, 1989.

[107] AZÁROFF L V, BROPHY J J. Electronic processes in materials[M]. New York: McGraw-Hill, 1963.

[108] GRIGSBY D L, JOHNSON D H, NEUBERGER M, et al. Electronic Properties of Materials[M]. Berlin: Springer-Verlag, 1993.